Interactive Student Edition

Reveal
MATH®

Course 3 • Volume 1

Mc
Graw
Hill

mheducation.com/prek-12

Send all inquiries to:
McGraw-Hill Education
STEM Learning Solutions Center
8787 Orion Place
Columbus, OH 43240

ISBN: 978-0-07-667375-9
MHID: 0-07-667375-8

Reveal Math, Course 3
Interactive Student Edition, Volume 1

Printed in the United States of America.

10 11 12 13 LMN 28 27 26 25 24 23 22 21

Contents in Brief

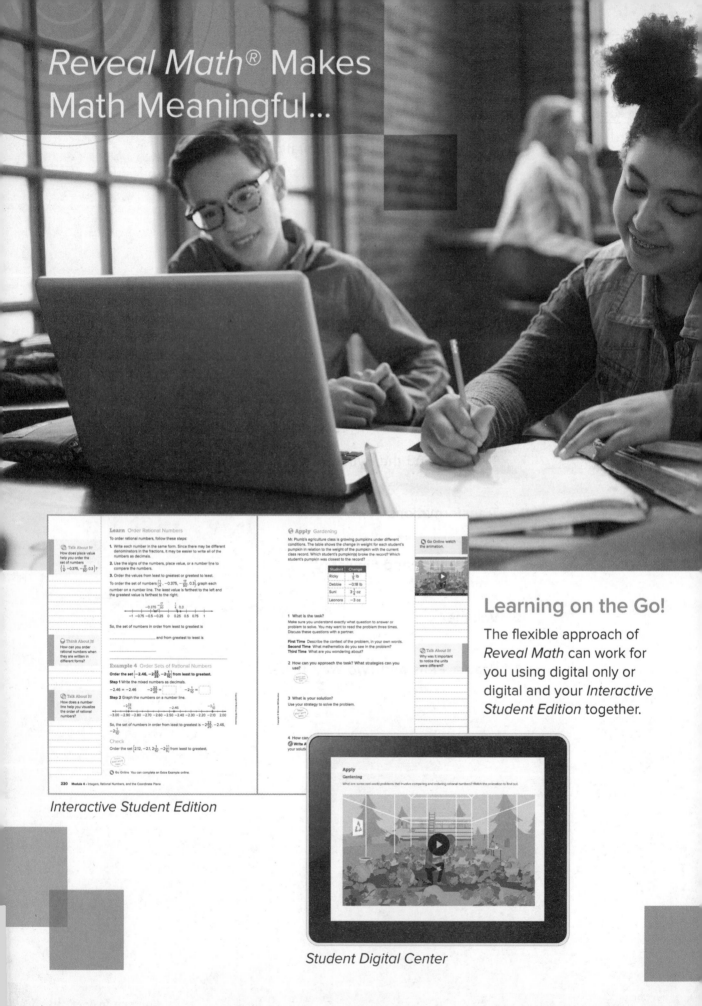

Reveal Math® Makes Math Meaningful...

Interactive Student Edition

Learn Order Rational Numbers

To order rational numbers, follow these steps:

1. Write each number in the same form. Since there may be different denominators in the fractions, it may be easier to write all of the numbers as decimals.

2. Use the signs of the numbers, place value, or a number line to compare the numbers.

3. Order the values from least to greatest or greatest to least.

Example 4 Order Sets of Rational Numbers

Apply Gardening

Mr. Plumb's agriculture class is growing pumpkins under different conditions. The table shows the change in weight for each student's pumpkin in relation to the weight of the pumpkin with the current class record. Which student's pumpkin(s) broke the record? Which student's pumpkin was closest to the record?

Student	Change
Ricky	$\frac{1}{8}$ lb
Debbie	-0.18 lb
Suni	$3\frac{1}{4}$ oz
Leonora	-3 oz

Student Digital Center

Apply
Gardening

Learning on the Go!

The flexible approach of *Reveal Math* can work for you using digital only or digital and your *Interactive Student Edition* together.

...to Reveal YOUR Full Potential!

Reveal Math® Brings Math to Life in Every Lesson

Reveal Math is a blended print and digital program that supports access on the go. You'll find the *Interactive Student Edition* aligns to the Student Digital Center, so you can record your digital observations in class and reference your notes later, or access just the digital center, or a combination of both! The Student Digital Center provides access to the interactive lessons, interactive content, animations, videos, and technology-enhanced practice questions.

Write down your username and password here

Username: _____

Password: _____

Go Online!
my.mheducation.com

Web Sketchpad® Powered by The Geometer's Sketchpad®- Dynamic, exploratory, visual activities embedded at point of use within the lesson.

Animations and Videos – Learn by seeing mathematics in action.

Interactive Tools – Get involved in the content by dragging and dropping, selecting, and completing tables.

Personal Tutors – See and hear a teacher explain how to solve problems.

eTools – Math tools are available to help you solve problems and develop concepts.

Module 1
Exponents and Scientific Notation

e Essential Question
Why are exponents useful when working with very large or very small numbers?

Module 2
Real Numbers

e Essential Question
Why do we classify numbers?

Module 3
Solve Equations with Variables on Each Side

℮ Essential Question

How can equations with variables on each side be used to represent everyday situations?

Module 4

Linear Relationships and Slope

ⓔ Essential Question

How are linear relationships related to proportional relationships?

TABLE OF CONTENTS

Module 5
Functions

e Essential Question
What does it mean for a relationship to be a function?

Module 6

Systems of Linear Equations

℮ Essential Question

How can systems of equations be helpful in solving everyday problems?

Module 7

Triangles and the Pythagorean Theorem

e Essential Question

How can angle relationships and right triangles be used to solve everyday problems?

Module 8
Transformations

ⓔ Essential Question

What does it mean to perform a transformation on a figure?

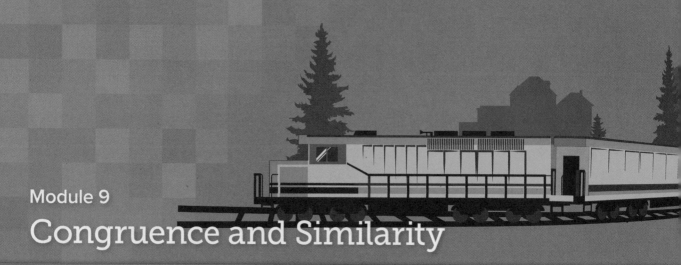

Module 9
Congruence and Similarity

e Essential Question
What information is needed to determine if two figures are congruent or similar?

Module 10
Volume

e Essential Question
How can you measure a cylinder, cone, or sphere?

Module 11
Scatter Plots and Two-Way Tables

e Essential Question

What do patterns in data mean and how are they used?

Module 1
Exponents and Scientific Notation

e Essential Question
Why are exponents useful when working with very large or very small numbers?

What Will You Learn?

Place a checkmark (✓) in each row that corresponds with how much you already know about each topic **before** starting this module.

KEY	Before			After		
⬛ — I don't know. ◆ — I've heard of it. ★ — I know it!	⬛	◆	★	⬛	◆	★
writing numerical products as powers						
evaluating powers						
multiplying numerical and algebraic powers						
dividing numerical and algebraic powers						
finding powers of numerical and algebraic powers						
finding powers of numerical and algebraic products						
simplifying expressions with negative exponents and zero exponents						
writing numbers in scientific notation						
computing with numbers in scientific notation						

📖 Foldables Cut out the Foldable and tape it to the Module Review at the end of the module. You can use the Foldable throughout the module as you learn about exponents and scientific notation.

What Vocabulary Will You Learn?

Check the box next to each vocabulary term that you may already know.

☐ base

☐ evaluate

☐ exponent

☐ monomial

☐ negative exponent

☐ order of operations

☐ Product of Powers Property

☐ power

☐ Power of a Power Property

☐ Power of a Product Property

☐ Quotient of Powers Property

☐ scientific notation

☐ standard form

☐ term

☐ Zero Exponent Rule

Are You Ready?

Study the Quick Review to see if you are ready to start this module. Then complete the Quick Check.

Quick Review	
Example 1 **Multiply integers.** Find $3 \cdot 2 \cdot 3 \cdot 2 \cdot 2$. $3 \cdot 2 \cdot 3 \cdot 2 \cdot 2 = 3 \cdot 3 \cdot 2 \cdot 2 \cdot 2$ $ = (3 \cdot 3) \cdot (2 \cdot 2 \cdot 2)$ $ = 9 \cdot 8$ $ = 72$	**Example 2** **Multiply rational numbers.** Find $2.8 \cdot 2.8$. $\quad 2.8 \quad \longleftarrow$ one decimal place $\times \ 2.8 \quad \longleftarrow$ one decimal place $\overline{\ 7.84} \quad \longleftarrow$ two decimal places

Quick Check	
1. A coach watched game film for $4 \cdot 2 \cdot 4 \cdot 4 \cdot 2$ hours last season. How many hours did the coach watch game film?	**2.** Find $(-1.3)(-1.3)(-1.3)$.

How Did You Do?

Which exercises did you answer correctly in the Quick Check? Shade those exercise numbers at the right.

Powers and Exponents

I Can... use integer exponents to show repeated multiplication of rational numbers.

Explore Exponents

Online Activity You will explore how to write repeated multiplication using exponents.

Emily is doubling the amount she saves each week, then adds this to her total savings. Complete the table to find these values. The first two columns have been completed for you.

Week	0	1	2	3	4	5
Weekly Savings	1¢	2¢	☐¢	☐¢	☐¢	☐¢
Total Savings	1¢	3¢	☐¢	☐¢	☐¢	☐¢

Clear All   Check Answer

Talk About It!

Learn Write Products as Powers

A product of repeated factors can be expressed as a **power**, that is, using an **exponent** and a **base**.

Go Online Watch the animation to learn how to write an expression as a power.

The animation shows that to write $3 \cdot 3 \cdot 3 \cdot 3$ as a power, express the number of times 3 is used as a factor as an exponent.

$$\overbrace{3 \cdot 3 \cdot 3 \cdot 3}^{\square \text{ factors}} = 3^4$$

The base, 3, is the common factor that is being multiplied.

The exponent, 4, tells how many times the base is used as a factor.

(continued on next page)

The expression $2 \cdot 2 \cdot 2 \cdot 2 \cdot (-4) \cdot (-4) \cdot (-4)$ has two different bases. To write this expression using exponents, express the number of times the base, 2, is used as a factor. Then express the number of times -4 is used as a factor.

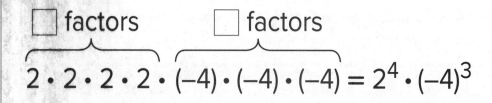

Label each part of the expression with the correct term.

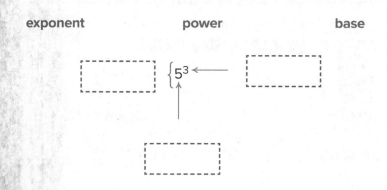

Complete the following statements about how powers are read.

$3^1 = 3$ 3 to the _____ power

$3^2 = 3 \cdot 3$ 3 to the _____ power or 3 _____

$3^3 = 3 \cdot 3 \cdot 3$ 3 to the _____ power or 3 _____

$3^4 = 3 \cdot 3 \cdot 3 \cdot 3$ 3 to the _____ power or 3 _____

$$\overbrace{3^n = 3 \cdot 3 \cdot 3 \cdot ... \cdot 3}^{n \text{ factors}}$$ 3 to the _____ power or 3 to the _____

Example 1 Write Numerical Products as Powers

Write the expression $(-2) \cdot (-2) \cdot (-2) \cdot \frac{3}{4} \cdot \frac{3}{4} \cdot \frac{3}{4} \cdot \frac{3}{4}$ **using exponents.**

The base -2 is used as a factor _____ times, and the base $\frac{3}{4}$ is

used as a factor _____ times. So, the exponents are _____

and _____.

So, $(-2) \cdot (-2) \cdot (-2) \cdot \frac{3}{4} \cdot \frac{3}{4} \cdot \frac{3}{4} \cdot \frac{3}{4} = (-2)^3 \cdot \left(\frac{3}{4}\right)^4$.

Check

Which of the following is equivalent to the expression

$\frac{2}{3} \cdot \frac{2}{3} \cdot \frac{2}{3} \cdot \frac{2}{3} \cdot (-5) \cdot (-5) \cdot (-5) \cdot (-5) \cdot (-5)$?

Ⓐ $\left(\frac{2}{3}\right)^3 \cdot (-5)^3$

Ⓑ $\left(\frac{2}{3}\right)^4 \cdot (-5)^3$

Ⓒ $\left(\frac{2}{3}\right)^4 \cdot (-5)^5$

Ⓓ $\left(\frac{2}{3}\right)^5 \cdot (-5)^4$

Show your work here

 Go Online You can complete an Extra Example online.

Pause and Reflect

Compare and contrast a product of repeated factors with its equivalent power.

Record your observations here

Talk About It!

Describe an advantage of writing a product of repeated factors as a power.

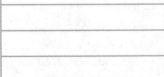

Example 2 Write Algebraic Products as Powers

Write the expression $a \cdot b \cdot b \cdot a \cdot b$ using exponents.

$$a \cdot b \cdot b \cdot a \cdot b = a \cdot a \cdot b \cdot b \cdot b \quad \text{Commutative Property}$$
$$= a^2 \cdot b^3 \qquad\qquad \text{The base } a \text{ is a factor 2 times, and base } b \text{ is a factor 3 times.}$$

So, $a \cdot b \cdot b \cdot a \cdot b = $ _____.

> **Talk About It!**
> In the first step, why was the Commutative Property used?

Check

Write the expression $a \cdot a \cdot b \cdot a \cdot b \cdot a \cdot a \cdot b \cdot b$ using exponents.

(A) $a^4 \cdot b^5$

(C) $a^5 \cdot b^4$

(B) $a^5 \cdot b^3$

(D) $a^6 \cdot b^4$

Show your work here

Go Online You can complete an Extra Example online.

Learn Negative Bases and Parentheses

For expressions that contain negative signs and/or parentheses, the inclusion and placement of parentheses can result in distinct expressions that have different values. For example, do you think that $(-a)^b$ and $-a^b$ have the same value?

Complete the following which compares and contrasts the expressions $(-a)^b$ and $-a^b$.

> **Talk About It!**
> When you evaluate $(-3)^4$ and -3^4, their results are opposites. When you evaluate $(-5)^3$ and -5^3, are the results opposites? Explain why or why not.

	$(-a)^b$	$-a^b$
Words	The expression $(-a)^b$ indicates that $-a$ is used as a factor b times.	The expression $-a^b$ means the opposite of a^b.
Variables	$(-a)^b = \underbrace{(-a) \cdot (-a) \cdot \ldots \cdot (-a)}_{b \text{ times}}$	$-a^b = -\underbrace{(a \cdot a \cdot \ldots \cdot a)}_{b \text{ times}}$
Numbers	$(-3)^4 = (-3) \cdot (-3) \cdot (-3) \cdot (-3)$ $= \boxed{}$	$-3^4 = -(3 \cdot 3 \cdot 3 \cdot 3)$ $= \boxed{}$

Learn Evaluate Powers

To **evaluate** an expression with a power, write the power as a product and then multiply.

When evaluating expressions with powers or more than one operation, it is important to remember to use the order of operations.

Order of Operations

1. Simplify the expression inside the grouping symbols.

2. Evaluate all powers.

3. Perform multiplication and division in order from left to right.

4. Perform addition and subtraction in order from left to right.

Example 3 Evaluate Numerical Expressions

Evaluate $(-2)^3 + (3.5)^2$.

$(-2)^3 + (3.5)^2 = (-2) \cdot (-2) \cdot (-2) + (3.5) \cdot (3.5)$ Write powers as products.

$$= \boxed{} + \boxed{}$$ Multiply.

$$= \boxed{}$$ Add.

So, $(-2)^3 + (3.5)^2 = 4.25$.

Check

Evaluate $8^3 + (-3)^3$.

 Go Online You can complete an Extra Example online.

Think About It!

According to the order of operations, what are the first evaluations you will need to do?

Talk About It!

Are the expressions $(-2)^3 + 5^2$ and $-2^3 + 5^2$ equivalent? Justify your response.

Example 4 Evaluate Algebraic Expressions

Evaluate $a^2 + b^4$ if $a = 3$ and $b = \frac{1}{2}$.

$$a^2 + b^4 = 3^2 + \left(\frac{1}{2}\right)^4 \qquad \text{Replace } a \text{ with 3 and } b \text{ with } \frac{1}{2}.$$

$$= (3 \cdot 3) + \left(\frac{1}{2} \cdot \frac{1}{2} \cdot \frac{1}{2} \cdot \frac{1}{2}\right) \qquad \text{Write the powers as products.}$$

$$= \boxed{} + \frac{\boxed{}}{\boxed{}} \qquad \text{Multiply.}$$

$$= \boxed{} \qquad \text{Add.}$$

So, $a^2 + b^4$ when $a = 3$ and $b = \frac{1}{2}$, is $9\frac{1}{16}$.

Check

Evaluate $a^4 + b^2$ if $a = -3$ and $b = 6$.

Show your work here

Example 5 Evaluate Algebraic Expressions

Evaluate $d^3 + (c^2 - 2)$ if $c = -4$ and $d = \frac{2}{5}$.

$$d^3 + (c^2 - 2) = \left(\frac{2}{5}\right)^3 + [(-4)^2 - 2] \qquad \begin{array}{l}\text{Replace } c \text{ with } -4 \text{ and } d \\ \text{with } \frac{2}{5}.\end{array}$$

$$= \left(\frac{2}{5}\right)^3 + \boxed{} \qquad \begin{array}{l}\text{Perform the operations in} \\ \text{grouping symbols.}\end{array}$$

$$= \boxed{} \qquad \text{Simplify.}$$

So, $d^3 + (c^2 - 2)$ when $c = -4$ and $d = \frac{2}{5}$, is $14\frac{8}{125}$.

Check

Evaluate $x^2 + (y^3 - 85)$ if $x = -\frac{1}{3}$ and $y = 5$.

Show your work here

🔵 **Go Online** You can complete an Extra Example online.

Think About It!

Before you can evaluate the expressions, what do you need to do?

Talk About It!

A classmate evaluated the expression $\left(\frac{1}{2}\right)^4$ by writing $\frac{1^4}{2}$ and simplifying it to $\frac{1}{2}$. Find and correct the error.

Talk About It!

When you replace c with -4, why should you evaluate $(-4)^2$ and not -4^2?

🌐 Apply Mammals

The table shows the average weights of two endangered mammals. How much more does the brown bear weigh than the panther?

Animal	Weight (lb)
Panther	$2^3 \cdot 3 \cdot 5$
Brown Bear	$2 \cdot 5^2 \cdot 7$

1 What is the task?

Make sure you understand exactly what question to answer or problem to solve. You may want to read the problem three times. Discuss these questions with a partner.

First Time Describe the context of the problem, in your own words.
Second Time What mathematics do you see in the problem?
Third Time What are you wondering about?

2 How can you approach the task? What strategies can you use?

3 What is your solution?

Use your strategy to solve the problem.

4 How can you show your solution is reasonable?

✏️ **Write About It!** Write an argument that can be used to defend your solution.

> 🗣 Talk About It!
> A female gorilla weighs $2^3 \cdot 5^2$ pounds. How does this compare to the panther's and brown bear's average weight?

Check

Interstate 90 stretches almost $2^4 \cdot 7^2 \cdot 4$ miles across the United States. Interstate 70 stretches almost $2^3 \cdot 5^2 \cdot 11$ miles across the United States. How much longer is Interstate 90 than Interstate 70?

Show your work here

Go Online You can complete an Extra Example online.

Pause and Reflect

How well do you understand the concepts from today's lesson? What questions do you still have? How can you get those questions answered?

Record your observations here

Practice

Go Online You can complete your homework online.

Write each expression using exponents. (Examples 1 and 2)

1. $(-7) \cdot (-7) \cdot 5 \cdot 5 \cdot 5 \cdot 5 =$ _____

2. $n \cdot n \cdot p \cdot p \cdot r \cdot r \cdot r =$ _____

Evaluate each numerical expression. (Example 3)

3. $3^4 - (-4)^2 =$ _____

4. $6 + 2^6 =$ _____

5. Evaluate $x^3 - y^2$ if $x = 2$ and $y = \frac{3}{4}$.
(Example 4)

6. Evaluate $(g + h)^3$ if $g = 2$ and $h = -3$.
(Example 5)

7. Replace $\square$ with $<$, $>$, or $=$ to make a true statement: $(-3)^4 \square (-4)^3$.

Test Practice

8. A scientist estimates that, after a certain amount of time, there would be $2^5 \cdot 3^3 \cdot 10^5$ bacteria in a Petri dish. How many bacteria is this?

9. **Multiselect** Select all of the expressions that evaluate to negative rational numbers.

☐ $(-9)^4$

☐ $\left(-\frac{4}{5}\right)^3$

☐ $3^5 - 10^4$

☐ $(9.8)^2 - 10^2$

☐ $\left(-\frac{3}{8}\right)^2$

Apply

10. The table shows the approximate number of species of each type of tree. How many more species of palm tree are there than maple tree?

Tree Type	Number of Species
Palm	$2^3 \cdot 3 \cdot 5^3$
Maple	2^7

11. The table shows the approximate number of lakes in two different states. How many more lakes does Florida have than Minnesota?

State	Number of Lakes
Florida	$3 \cdot 10^4$
Minnesota	$2^3 \cdot 5^2 \cdot 109$

12. ⓂⓅ **Identify Structure** Without evaluating, explain why $(-8.4)^5$ is less than 2^2.

13. ⓂⓅ **Find the Error** A student evaluated the expression $(x^2 - y)^3$ if $x = -4$ and $y = 7$. Find his mistake and correct it.

$$(x^2 - y)^3 = (-4^2 - 7)^3$$
$$= (-16 - 7)^3$$
$$= (-23)^3$$
$$= -12{,}167$$

14. Determine if the statement is *true* or *false*. Justify your response.

The expressions $(-x)^y$ and $-x^y$ have the same value.

15. A student finds that $\left(\frac{1}{4}\right)^2 = \left(\frac{1}{2}\right)^4$ and concludes, then, that $\left(\frac{1}{3}\right)^2 = \left(\frac{1}{2}\right)^3$. Is this reasoning correct? Explain.

Multiply and Divide Monomials

I Can... use the Laws of Exponents to multiply and divide monomials with common bases.

Explore Product of Powers

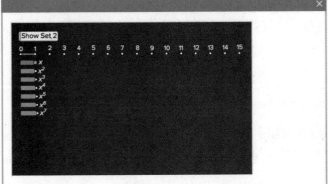 **Online Activity** You will use Web Sketchpad to explore how to simplify a product of powers with like bases.

What Vocabulary Will You Learn?

monomial

Product of Powers Property

Quotient of Powers Property

term

Learn Monomials

A **monomial** is a number, a variable, or a product of a number and one or more variables. For example, $6y^2$ is a monomial since it is a product of 6 and y^2. The expression $x + 3$ is not a monomial since it is a sum of two monomials.

When addition or subtraction signs separate an algebraic expression into parts, each part is called a **term**. A monomial only has one term. The expression $x + 3$ is a sum of two monomials and therefore not a monomial itself because it has two terms.

Write each expression in the appropriate bin. An example of each type is given.

x 80 $x^2 - y^2$ $8x$ $x + 5$

Monomials

$6y^2$

Not Monomials

$x + 3$

Learn Product of Powers

🅚 **Go Online** Watch the animation to simplify the product of powers with the same base.

For example, to simplify $4^2 \cdot 4^3$, the animation shows to:

Step 1 Write 4^2 as a product of 2 factors.

Step 2 Write 4^3 as a product of 3 factors.

$$4^2 \cdot 4^3 = \underbrace{\overbrace{(4 \cdot 4)}^{2 \text{ factors}} \cdot \overbrace{(4 \cdot 4 \cdot 4)}^{3 \text{ factors}}}_{5 \text{ factors}}$$

There are 5 common factors all together, so the product is 4^5.

Notice that the sum of the exponents of the original powers is the exponent in the final product.

$$4^2 \cdot 4^3 = 4^{2+3}$$
$$= 4^5$$

You can multiply powers with the same base by adding the exponents.

You can simplify a product of powers with like bases using the **Product of Powers Property**. The group of integer exponent properties, including the Product of Powers Property, is called the Laws of Exponents.

💬 **Talk About It!**

When simplifying a product of powers using the Product of Powers Property, why do the bases have to be the same? For example, why can't you use the Product of Powers Property to simplify $x^3 \cdot y^2$?

Words	Algebra
To multiply powers with the same base, add their exponents.	$a^m \cdot a^n = a^{m+n}$
	Numbers
	$2^4 \cdot 2^3 = 2^{4+3}$ or ⬚

Example 1 Multiply Numerical Powers

Simplify $5^4 \cdot 5$.

$$5^4 \cdot 5 = 5^4 \cdot 5^1 \qquad\qquad 5 = 5^1$$
$$= 5^{4+1} \qquad\qquad \text{Product of Powers Property}$$
$$= \boxed{} \text{ or } \boxed{} \qquad \text{Add the exponents. Simplify.}$$

So, $5^4 \cdot 5 = 5^5$ or 3,125.

Check

Simplify the expression $4^5 \cdot 4^3$.

Show your work here

Think About It!

How will the Product of Powers Property help you simplify the expression?

Talk About It!

Describe another method you can use to simplify the expression.

Example 2 Multiply Algebraic Powers

Simplify $c^3 \cdot c^5$.

$$c^3 \cdot c^5 = c^{3+5} \qquad\qquad \text{Product of Powers Property}$$
$$= \boxed{} \qquad\qquad\qquad \text{Simplify.}$$

So, $c^3 \cdot c^5 = c^8$.

Check

Simplify the expression $x^4 \cdot x^6$.

Show your work here

Talk About It!

Explain why you were able to add the exponents to simplify the expression $c^3 \cdot c^5$.

Go Online You can complete an Extra Example online.

Think About It!

When you multiply the two monomials, what will you do with the coefficients?

Talk About It!

When simplifying the expression, why were the coefficients multiplied but the exponents added?

Example 3 Multiply Monomials

Simplify $-3x^2 \cdot 4x^5$.

$$-3x^2 \cdot 4x^5 = -3 \cdot x^2 \cdot 4 \cdot x^5 \qquad \text{Definition of coefficient}$$

$$= (-3 \cdot 4)(x^2 \cdot x^5) \qquad \text{Commutative and Associative Property}$$

$$= \boxed{}(x^2 \cdot x^5) \qquad \text{Multiply the coefficients.}$$

$$= -12(x^{2+5}) \qquad \text{Product of Powers Property}$$

$$= \boxed{} \qquad \text{Simplify.}$$

So, $-3x^2 \cdot 4x^5 = -12x^7$.

Check

Simplify the expression $-2a(3a^4)$.

(A) $1a^5$ (C) $-6a^4$

(B) $6a^5$ (D) $-6a^5$

Show your work here

Go Online You can complete an Extra Example online.

Explore Quotient of Powers

Online Activity You will use Web Sketchpad to explore how to simplify a quotient of powers with like bases.

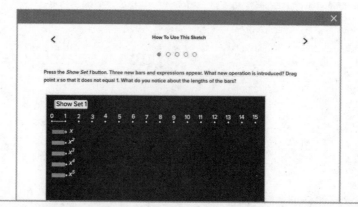

Learn Quotient of Powers

Go Online Watch the animation to learn how to simplify the quotient of powers with the same base.

For example, to simplify $\frac{3^6}{3^2}$, the animation shows to:

Step 1 Write 3^6 as a product of 6 factors.

Step 2 Write 3^2 as a product of 2 factors.

Step 3 Divide out common factors and simplify.

$$\frac{3^6}{3^2} = \frac{\overbrace{3 \cdot 3 \cdot 3 \cdot 3 \cdot \cancel{3} \cdot \cancel{3}}^{6 \text{ factors}}}{\underbrace{\cancel{3} \cdot \cancel{3}}_{2 \text{ factors}}} = \frac{3 \cdot 3 \cdot 3 \cdot 3}{1} \text{ or } 3^4$$

Four factors of 3 remain in the numerator and 1 in the denominator, so the quotient is 3^4.

Notice that the difference of the exponents of the original powers is the exponent in the final quotient.

$$\frac{3^6}{3^2} = 3^{6-2}$$
$$= 3^4$$

You can divide powers with the same base by subtracting the exponents.

You can simplify a quotient of powers with like bases using the **Quotient of Powers Property**.

Words	Algebra
To divide powers with the same base, subtract their exponents.	$\frac{a^m}{a^n} = a^{m-n}$, where $a \neq 0$.
	Numbers
	$\frac{3^7}{3^3} = 3^{7-3}$ or $\boxed{}$

> **Talk About It!**
> When simplifying a quotient of powers using the Quotient of Powers Property, why do the bases have to be the same? For example, why can't you use the Quotient of Powers Property to simplify $\frac{x^8}{y^3}$?

Example 4 Divide Algebraic Powers

Simplify $\frac{x^8}{x^2}$.

$$\frac{x^8}{x^2} = x^{8-2}$$ Quotient of Powers Property

$$= \boxed{}$$ Subtract the exponents.

So, $\frac{x^8}{x^2} = x^6$.

Check

Simplify $\frac{z^{12}}{z^4}$.

 Go Online You can complete an Extra Example online.

Example 5 Divide Powers

Hawaii's total shoreline is about 2^{10} miles long. New Hampshire's shoreline is about 2^7 miles long.

About how many times longer is Hawaii's shoreline than New Hampshire's?

To determine how many times longer Hawaii's shoreline is than New Hampshire's, divide their lengths.

$$\frac{2^{10}}{2^7} = 2^{10-7}$$ Quotient of Powers Property

$$= \boxed{}$$ Subtract the exponents.

$$= \boxed{}$$ Simplify.

So, Hawaii's shoreline is about 8 times longer than New Hampshire's shoreline.

Talk About It!

Describe another method you could use to simplify the expression.

Think About It!

What operation will you perform to find how many times longer one shoreline is than the other?

Talk About It!

Alaska's shoreline is about 3^8 miles long. Explain why you could not use the Quotient of Powers Property to simplify the expression $\frac{3^8}{2^7}$ to find out how many times longer Alaska's shoreline is than New Hampshire's.

Check

The table shows the seating capacity of two different facilities. About how many times as great is the capacity of Madison Square Garden in New York City than a typical movie theater?

Place	Seating Capacity
Movie Theater	3^5
Madison Square Garden	3^9

Ⓐ 3^4 or 64 times

Ⓑ 3^4 or 81 times

Ⓒ 3^5 or 243 times

Ⓓ 3^{14} or 4,782,969 times

Show your work here

 Go Online You can complete an Extra Example online.

Example 6 Divide Numerical Powers

Simplify $\dfrac{2^5 \cdot 3^5 \cdot 5^2}{2^2 \cdot 3^4 \cdot 5}$.

$$\frac{2^5 \cdot 3^5 \cdot 5^2}{2^2 \cdot 3^4 \cdot 5} = \left(\frac{2^5}{2^2}\right)\left(\frac{3^5}{3^4}\right)\left(\frac{5^2}{5}\right)$$ Group by common base.

$$= 2^{5-2} \cdot 3^{5-4} \cdot 5^{2-1}$$ Quotient of Powers Property

$$= \boxed{} \cdot \boxed{} \cdot \boxed{}$$ Subtract the exponents.

$$= \boxed{} \cdot \boxed{} \cdot \boxed{}$$ Evaluate the powers.

$$= \boxed{}$$ Simplify.

So, $\dfrac{2^5 \cdot 3^5 \cdot 5^2}{2^2 \cdot 3^4 \cdot 5} = 120$.

Talk About It!

Why is it important to group terms by their common base(s)?

Check

Simplify the expression $\dfrac{2^2 \cdot 3^3 \cdot 4^5}{2 \cdot 3 \cdot 4^4}$.

Ⓐ $2 \cdot 3^2 \cdot 4$

Ⓑ $2 \cdot 3^3 \cdot 4$

Ⓒ $2^2 \cdot 3^2 \cdot 4$

Ⓓ $2^2 \cdot 3^3 \cdot 4$

Show your work here

🍄 **Think About It!**

How can you begin solving the problem?

Example 7 Divide Monomials

Simplify $\dfrac{12w^5}{2w}$.

$$\dfrac{12w^5}{2w} = \left(\dfrac{12}{2}\right)\left(\dfrac{w^5}{w}\right) \qquad \text{Associative Property}$$

$$= \boxed{}\left(\dfrac{w^5}{w}\right) \qquad \text{Divide the coefficients.}$$

$$= 6\,(w^{5-1}) \qquad \text{Quotient of Powers Property}$$

$$= \boxed{} \qquad \text{Subtract the exponents.}$$

So, $\dfrac{12w^5}{2w} = 6w^4$.

💬 **Talk About It!**

Why were the coefficients divided, but the exponents subtracted?

Check

Simplify the expression $\dfrac{24k^9}{6k^5}$.

Show your work here

🔎 **Go Online** You can complete an Extra Example online.

⊕ Apply Computer Science

The processing speed of a certain computer is 10^{12} instructions per second. A second computer has a processing speed that is 10^5 times as fast as the first computer. A third computer has a processing speed of 10^{15} instructions per second. Which computer has the fastest processing speed?

1 What is the task?

Make sure you understand exactly what question to answer or problem to solve. You may want to read the problem three times. Discuss these questions with a partner.

First Time Describe the context of the problem, in your own words.
Second Time What mathematics do you see in the problem?
Third Time What are you wondering about?

2 How can you approach the task? What strategies can you use?

Record your observations here

3 What is your solution?

Use your strategy to solve the problem.

Show your work here

🗨 **Talk About It!**

How did understanding the Laws of Exponents help you solve the problem?

4 How can you show your solution is reasonable?

◢ **Write About It!** Write an argument that can be used to defend your solution.

Check

Bolivia has a population of 10^7 people, while the population of Aruba is 10^5 people. The Philippines has a population that is 10^3 times greater than Aruba's population. Which statement is true about the population of the three countries?

(A) Aruba has the least population.

(B) Boliva has the greatest population.

(C) The Phillipines has the least population.

(D) Bolivia and the Phillipines have the same population.

Show your work here

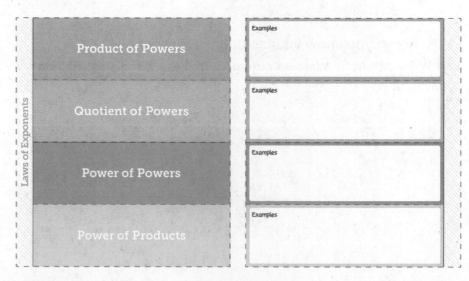

Go Online You can complete an Extra Example online.

Foldables It's time to update your Foldable, located in the Module Review, based on what you learned in this lesson. If you haven't already assembled your Foldable, you can find the instructions on page FL1.

Laws of Exponents		Examples
Product of Powers		Examples
Quotient of Powers		Examples
Power of Powers		Examples
Power of Products		

Practice

Go Online You can complete your homework online.

Simplify each expression. (Examples 1–3)

1. $3^8 \cdot 3 =$ _____

2. $m^5 \cdot m^2 =$ _____

3. $3m^3n^2 \cdot 8mn^3 =$ _____

4. $9p^4 \cdot (-8p^2) =$ _____

5. Simplify $\dfrac{b^{12}}{b^5}$. (Example 4)

6. Simplify $\dfrac{5^5 \cdot 6^3 \cdot 8^{10}}{5^3 \cdot 6 \cdot 8^9}$. (Example 6)

7. A publisher sells 10^6 copies of a new science fiction book and 10^3 copies of a new mystery book. How many times as many science fiction books were sold than mystery books? (Example 5)

Test Practice

8. Simplify $\dfrac{45x^{15}}{9x^{10}}$. (Example 7)

9. Equation Editor Simplify $\dfrac{a^4c^6}{a^2c}$.

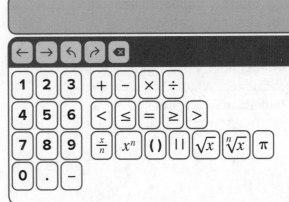

Apply

10. Fouster's Farms has 5^3 fruit trees on their land. Myrna's Farms has five times as many fruit trees as Fouster's. A commercial farm has 500 fruit trees. Which farm has the most fruit trees?

11. The table shows the number of bacteria in each Petri dish. Dish C has 8^2 times as many bacteria has Dish A. Which Petri dish holds the most number of bacteria?

Petri Dish	Number of Bacteria
A	8^6
B	8^9

12. Write a multiplication expression with a product of 8^{13}.

13. 🔲 **Persevere with Problems** What is four times 4^{15}? Write using exponents and explain your reasoning.

14. 🔲 **Identify Repeated Reasoning** Consider the sequence below:

    2, 4, 8, 16, 32, 64, ...

    The number 4,096 belongs to this sequence. What is the number that immediately precedes it?

15. What value of n makes a true statement?

    $$4^n \cdot 4^2 = 16{,}384$$

Powers of Monomials

I Can... use the Power of a Power Property and the Power of a Product Property to simplify expressions with integer exponents.

What Vocabulary Will You Learn?
Power of a Power Property

Power of a Product Property

Explore Power of a Power

Online Activity You will explore how to simplify a power raised to another power.

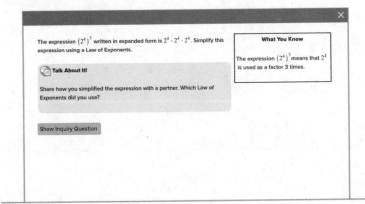

The expression $\left(2^{4}\right)^{3}$ written in expanded form is $2^{4} \cdot 2^{4} \cdot 2^{4}$. Simplify this expression using a Law of Exponents.

Talk About It!

Share how you simplified the expression with a partner. Which Law of Exponents did you use?

Show Inquiry Question

What You Know

The expression $\left(2^{4}\right)^{3}$ means that 2^{4} is used as a factor 3 times.

Learn Power of a Power

You can use the rule for finding the product of powers to illustrate how to find the power of a power.

$(6^4)^3 = (6^4)(6^4)(6^4)$ Expand the 3 factors.

$= 6^{4+4+4}$ Product of Powers Property

$= \boxed{}$ Add the exponents.

Notice, that the product of the original exponents, 4 and 3, is the final power, 12. You can simplify a power raised to another power using the **Power of a Power Property.**

Words	Algebra
To find the power of a power, multiply the exponents.	$(a^m)^n = a^{m \cdot n}$
	Numbers
	$(5^2)^3 = 5^{2 \cdot 3}$ or $\boxed{}$

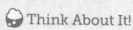
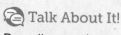
Example 1 Power of a Power

Simplify $(8^6)^3$.

$(8^6)^3 = 8^{6 \cdot 3}$ Power of a Power Property

     $= \boxed{}$ Simplify.

So, $(8^6)^3 = 8^{18}$.

Check

Simplify the expression $(6^3)^5$.

Example 2 Power of a Power

Simplify $(k^7)^5$.

$(k^7)^5 = k^{7 \cdot 5}$ Power of a Power Property

     $= \boxed{}$ Simplify.

So, $(k^7)^5 = k^{35}$.

Check

Simplify the expression $(x^4)^7$.

Go Online You can complete an Extra Example online.

Learn Power of a Product

The following example demonstrates how the Power of a Power Property can be extended to find the power of a product.

$(4a^2)^3 = (4a^2)(4a^2)(4a^2)$ Expand the 3 factors.

$\qquad = 4 \cdot 4 \cdot 4 \cdot a^2 \cdot a^2 \cdot a^2$ Commutative Property

$\qquad = 4^3 \cdot (a^2)^3$ Definition of power

$\qquad = 64 \cdot a^6$ or $64a^6$ Power of a Power Property

Notice, that each factor inside the parentheses is raised to the

_____ power.

You can simplify a product raised to a power using the **Power of a Product Property.**

Words	Algebra
To find the power of a product, find the power of each factor and multiply.	$(ab)^m = a^m b^m$
	Numbers
	$(2x^3)^4 = (2)^4 \cdot (x^3)^4$ or ☐

Example 3 Power of a Product

Simplify $(2p^3)^4$.

$(2p^3)^4 = 2^4 \cdot (p^3)^4$ Power of a Product Property

$\qquad = 2^4 \cdot p^{3 \cdot 4}$ Power of a Power Property

$\qquad = \boxed{}$ Simplify.

So, $(2p^3)^4 = 16p^{12}$.

Check

Simplify the expression $(7w^7)^3$.

Ⓐ $7w^{21}$

Ⓑ $21w^{21}$

Ⓒ $243w^{10}$

Ⓓ $343w^{21}$

Show your work here

 Go Online You can complete an Extra Example online.

Think About It!

What parts of the monomial, $2p^3$, are raised to the fourth power?

Talk About It!

Why was the exponent 4 applied to both of the factors in the first step?

What parts of the monomial, $(-2m^7n^6)$, are raised to the fifth power?

 Talk About It!

Why were the exponents multiplied when simplifying the expression?

Example 4 Power of a Product

Simplify $(-2m^7n^6)^5$.

$(-2m^7n^6)^5 = (-2)^5 \cdot (m^7)^5 \cdot (n^6)^5$ Power of a Product Property

$= \boxed{}$ Simplify.

So, $(-2m^7n^6)^5 = -32m^{35}n^{30}$.

Check

Simplify the expression $(-5w^2z^8)^3$.

(A) $-125w^5z^{11}$

(B) $-125w^6z^{24}$

(C) $-15w^5z^{11}$

(D) $15w^6z^{24}$

 Show your work here

 Go Online You can complete an Extra Example online.

Pause and Reflect

Did you ask questions about the properties used in this lesson? Why or why not?

Record your observations here

🌐 Apply Geometry

A square floor has a side length of $8x^3y^2$ units. A square tile has a side length of xy units. How many tiles will it take to cover the floor?

🅡 Go Online
Watch the animation.

1 What is the task?

Make sure you understand exactly what question to answer or problem to solve. You may want to read the problem three times. Discuss these questions with a partner.

First Time Describe the context of the problem, in your own words.
Second Time What mathematics do you see in the problem?
Third Time What are you wondering about?

2 How can you approach the task? What strategies can you use?

Record your observations here

3 What is your solution?

Use your strategy to solve the problem.

Show your work here

💬 Talk About It!
To find the number of tiles needed to cover the floor, why do you need to divide the areas?

4 How can you show your solution is reasonable?

✏️ **Write About It!** Write an argument that can be used to defend your solution.

Check

The floor of the commons room at King Middle School is in the shape of a square with side lengths of $3x^2y^3$ feet. New tile is going to be put on the floor of the room. The square tile has side lengths of xy feet. How many tiles will it take to cover the floor?

Show your work here

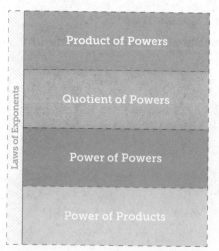

Go Online You can complete an Extra Example online.

Foldables It's time to update your Foldable, located in the Module Review, based on what you learned in this lesson. If you haven't already assembled your Foldable, you can find the instructions on page FL1.

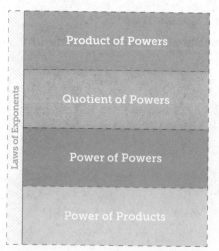

Laws of Exponents

Product of Powers

Quotient of Powers

Power of Powers

Power of Products

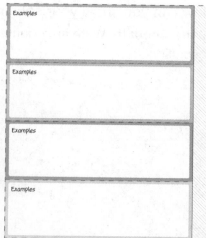

Examples

Examples

Examples

Examples

Practice

Go Online You can complete your homework online.

Simplify each expression. (Examples 1–4)

1. $(7^2)^3 = $ _____

2. $(8^3)^3 = $ _____

3. $(d^7)^6 = $ _____

4. $(z^7)^3 = $ _____

5. $(2m^5)^6 = $ _____

6. $(7a^5b^6)^4 = $ _____

7. $(-3w^3z^8)^5 = $ _____

8. $(-5r^4s^{10})^4 = $ _____

9. Which is greater: 1,000 or $(6^2)^3$? Explain.

Test Practice

10. Multiselect Select all of the expressions that simplify to the same expression.

☐ $(x^3y^4)^2$

☐ $(x^2y)^2$

☐ $(x^3)^2y^6$

☐ $x^6(y^4)^2$

☐ $(x^3)^2(y^2)^4$

Apply

11. A square floor has a side length of $7x^5y^6$ units. A square tile has a side length of x^2y units. How many tiles will it take to cover the floor?

12. A cube has a side length of $3x^6$ units. A smaller cube has a side length of x^2 units. How many smaller cubes will fit in the larger cube?

13. Make an argument for why $(4^2)^4 = (4^4)^2$.

14. 🆖 **Persevere with Problems** Describe all the positive integers that would make $\left[\left(\frac{1}{2}\right)^4\right]^n$ less than $\left(\frac{1}{2}\right)^7$. Explain your reasoning.

15. Without computing, determine which number is greater, $[(-4)^4]^5$ or $-[(4^{12})^3]$. Explain your reasoning.

16. 🆖 **Identify Structure** Charlotte states that $(4^3)^3$ can be rewritten as 2^{18}. Explain how she is correct.

Zero and Negative Exponents

I Can... use the Zero Exponent Rule and the Quotient of Powers Property to simplify expressions with zero and negative integer exponents.

What Vocabulary Will You Learn?
negative exponent
Zero Exponent Rule

Explore Exponents of Zero

Online Activity You will explore how to simplify expressions with exponents of zero and make a conjecture about the value of expressions with exponents of zero.

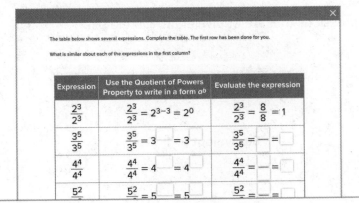

The table below shows several expressions. Complete the table. The first row has been done for you.

What is similar about each of the expressions in the first column?

Expression	Use the Quotient of Powers Property to write in a form a^b	Evaluate the expression
$\dfrac{2^3}{2^3}$	$\dfrac{2^3}{2^3} = 2^{3-3} = 2^0$	$\dfrac{2^3}{2^3} = \dfrac{8}{8} = 1$
$\dfrac{3^5}{3^5}$	$\dfrac{3^5}{3^5} = 3 = 3$	$\dfrac{3^5}{3^5} = \dfrac{}{} = \square$
$\dfrac{4^4}{4^4}$	$\dfrac{4^4}{4^4} = 4 = 4$	$\dfrac{4^4}{4^4} = \dfrac{}{} = \square$
$\dfrac{5^2}{}$	$\dfrac{5^2}{} = 5 = 5$	$\dfrac{5^2}{} = \dfrac{}{} = \square$

Learn Exponents of Zero

Use the **Zero Exponent Rule** to simplify expressions containing exponents of zero.

Words
Any nonzero number to the zero power is equivalent to 1.
Algebra
$x^0 = 1, x \neq 0$
Numbers
$5^0 = \square$

(continued on next page)

Complete the pattern in the table to demonstrate that any nonzero number, n, to the zero power is equivalent to 1.

Power	Equivalent Form
n^5	$n \cdot n \cdot n \cdot n \cdot n$
n^4	
n^3	
n^2	
n^1	
n^0	

n^{5-1} ... $\div n$
n^{4-1} ... $\div n$
n^{3-1} ... $\div n$
n^{2-1} ... $\div n$
n^{1-1} ... $\div n$

The pattern in the table shows that as you decrease the exponent by 1, the value of the power is divided by n each time. By extending the pattern, $n \div n = 1$. So, $n^0 = 1$.

Write the expressions that are equivalent to 1 in the bin.

$$2^1 \qquad 1^2 \qquad 2^0 \qquad 1^0 \qquad 0^1 \qquad \left(\frac{1}{2}\right)^0$$

Equivalent to 1

Pause and Reflect

Do all powers with any rational number base and an exponent of zero equal 1? Explain.

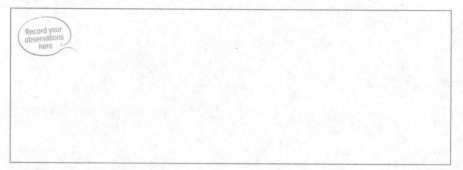

Record your observations here

Example 1 Exponents of Zero

Simplify 12^0.

Any nonzero number to the zero power is _____.

So, $12^0 = 1$.

Check

Simplify m^0, where $m \neq 0$.

Go Online You can complete an Extra Example online.

Explore Negative Exponents

Online Activity You will use Web Sketchpad to explore how to simplify expressions with negative exponents.

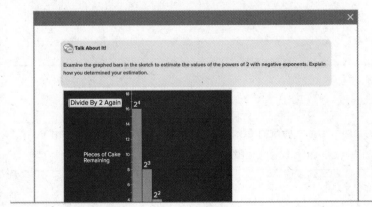

Learn Negative Exponents

A **negative exponent** is the result of repeated division. You can use negative exponents to represent very small numbers.

Complete the table below. Every time you divide by 10, the exponent decreases by one. The pattern in the table shows that 10^{-2} can be defined as $\frac{1}{100}$ or $\frac{1}{10^2}$.

Exponential Form	Standard Form
$10^3 = 10 \cdot 10 \cdot 10$	
$10^2 = 10 \cdot 10$	
10^1	
10^0	
10^{-1}	
10^{-2}	
10^{-3}	

÷ 10
÷ 10
÷ 10
÷ 10
÷ 10
÷ 10

Words
Any nonzero number to the negative n power is the multiplicative inverse of its nth power.
Algebra
$x^{-n} = \frac{1}{x^n}, x \neq 0$
Numbers
$7^{-3} = \frac{1}{7} \cdot \frac{1}{7} \cdot \frac{1}{7}$ or $\frac{1}{7^3}$

Pause and Reflect

Did you make any errors when completing the Learn? What can you do to make sure you don't repeat that error in the future?

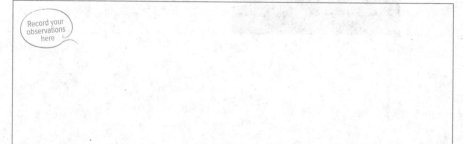

Record your observations here

Example 2 Negative Exponents

Express 6^{-3} using a positive exponent.

$6^{-3} = \dfrac{1}{\boxed{}}$ Definition of negative exponent

So, 6^{-3} expressed using a positive exponent is $\dfrac{1}{6^3}$.

Check

Express a^{-5} using a positive exponent.

Show your work here

Example 3 Negative Exponents

Express the fraction $\dfrac{1}{c^5}$ using a negative exponent.

$\dfrac{1}{c^5} = c^{-5}$ Definition of negative exponent

So, $\dfrac{1}{c^5}$ expressed using a negative exponent is _____.

Check

Express the fraction $\dfrac{1}{9^7}$ using a negative exponent.

Show your work here

Think About It!

What is the multiplicative inverse of 6?

Talk About It!

Explain how to write x^{-5} using a positive exponent.

Go Online You can complete an Extra Example online.

Think About It!

Will you simplify using the Product of Powers Property first, or will you simplify the negative exponent first?

Example 4 Negative Exponents

Simplify $5^3 \cdot 5^{-5}$.

An expression is in simplest form if it contains no like bases and no negative exponents.

$$5^3 \cdot 5^{-5} = 5^{3+(-5)}$$ Product of Powers Property

$$= \boxed{}$$ Simplify.

$$= \dfrac{1}{\boxed{}}$$ Write using positive exponents.

$$= \dfrac{1}{\boxed{}}$$ Simplify.

So, $5^3 \cdot 5^{-5} = \dfrac{1}{25}$.

Check

Simplify $3^9 \cdot 3^{-4}$.

 Talk About It!

Why is the answer not left as 5^{-2}?

Example 5 Negative Exponents

Simplify $\dfrac{w^{-1}}{w^{-4}}$.

$$\dfrac{w^{-1}}{w^{-4}} = w^{-1-(-4)}$$ Quotient of Powers Property

$$= \boxed{}$$ Simplify.

So, $\dfrac{w^{-1}}{w^{-4}} = w^3$.

Check

Simplify $\dfrac{b^{-7}}{b^{-4}}$.

 Talk About It!

Both of the original exponents are negative. Why is the exponent in the simplified expression positive?

 Go Online You can complete an Extra Example online.

🌐 Apply Measurement

One strand of human hair is about 0.001 inch in diameter. The diameter of a certain wire for jewelry is 10^{-2} inch. How many times larger is the diameter of the wire than a strand of hair? Write the decimal as a power of 10.

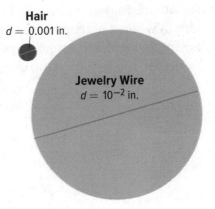

Hair
$d = 0.001$ in.

Jewelry Wire
$d = 10^{-2}$ in.

1 What is the task?

Make sure you understand exactly what question to answer or problem to solve. You may want to read the problem three times. Discuss these questions with a partner.

First Time Describe the context of the problem, in your own words.
Second Time What mathematics do you see in the problem?
Third Time What are you wondering about?

2 How can you approach the task? What strategies can you use?

Record your observations here

3 What is your solution?

Use your strategy to solve the problem.

Show your work here

4 How can you show your solution is reasonable?

✏️ **Write About It!** Write an argument that can be used to defend your solution.

💬 **Talk About It!**
Why might writing 0.001 as a fraction be helpful?

Check

An American green tree frog tadpole is about 0.00001 kilometer in length when it hatches. The largest indoor swimming pool is 10^{-1} kilometer wide. How many times longer is the width of the swimming pool than the length of the green tree frog tadpole?

Show your work here

Go Online You can complete an Extra Example online.

Pause and Reflect

Where did you encounter struggle in this lesson, and how did you deal with it? Write down any questions you still have.

Record your observations here

Practice

Go Online You can complete your homework online.

Simplify each expression. (Example 1)

1. $46^0 =$ _____

2. w^0, where $w \neq 0$

_____

Express each using a positive exponent.
(Example 2)

3. $8^{-4} =$ _____

4. $y^{-9} =$ _____

Express each fraction using a negative exponent. (Example 3)

5. $\dfrac{1}{d^6} =$ _____

6. $\dfrac{1}{10^5} =$ _____

Simplify each expression. (Examples 4 and 5)

7. $9^4 \cdot 9^{-6} =$ _____

8. $y^{-9} \cdot y^3 =$ _____

9. $\dfrac{x^{-8}}{x^{-12}} =$ _____

10. $\dfrac{d^{-13}}{d^{-2}} =$ _____

11. Simplify $8^{-7} \cdot 8^7 \cdot 10^4 \cdot 10^{-4}$.

Test Practice

12. Multiselect Select all of the expressions that are simplified.

☐ n^4

☐ $\dfrac{1}{n^{-5}}$

☐ $n^6 \cdot n^{-8}$

☐ $n^7 \cdot p^8$

☐ $\dfrac{1}{n^3}$

Apply

13. A dish containing bacteria has a diameter of 0.0001 kilometer. The diameter of a bacterium is 10^{-16} kilometer. How many times larger is the diameter of the dish than the diameter of the bacterium?

14. The table shows the diameters of two objects. How many times larger is the diameter of a pinhead than the diameter of a cloud water droplet?

Object	Diameter (m)
Cloud Water Droplet	10^{-5}
Pinhead	0.001

15. Without evaluating, order 5^{-7}, 5^4, and 5^0 from least to greatest. Explain your reasoning.

16. (MP) **Persevere with Problems** Explain how to find the value of $\left(\frac{1}{3^{-2}}\right)^{-3}$.

17. Determine if the following numerical expressions are equivalent. Explain your reasoning.

$$\left[\left(\frac{1}{10^{-2}}\right)^0\right]^3 \text{ and } [(10^{-4})^3]^0$$

18. Determine if the following statement is *true* or *false*. Explain your reasoning.

If a number between 0 and 1 is raised to a negative power, the result is a number greater than 1.

Scientific Notation

I Can... write very large and very small numbers using scientific notation.

What Vocabulary Will You Learn?

scientific notation

standard form

Explore Scientific Notation

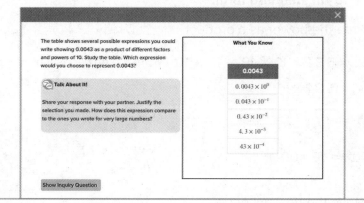

Online Activity You will explore how to write very large and very small numbers using scientific notation.

Learn Scientific Notation

Numbers that do not contain exponents are written in **standard form**. However, when you work with very large or very small numbers, it can be difficult to keep track of the place value. **Scientific notation** is a compact way of writing very large or very small numbers.

Words
Scientific notation is a way of expressing a number as the product of a factor and an integer power of 10. When the number is positive, the factor must be greater than or equal to 1 and less than 10.
Symbols
For positive numbers written in scientific notation, $a \times 10^n$, where $1 \le a < 10$ and n is an integer
Examples
$425{,}000{,}000 = 4.25 \times 10^8$ $0.00003 = 3 \times 10^{-5}$

(continued on next page)

Circle yes or no to determine whether each number is written in scientific notation.

10.6×10^{14}	Yes	No
2.019×10^{12}	Yes	No
9.5×10^{-7}	Yes	No
0.526×10^{-2}	Yes	No

Example 1 Write Numbers in Standard Form

Write 5.34×10^4 in standard form.

Method 1 Write the power as a product.

$$5.34 \times 10^4 = 5.34 \times 10 \times 10 \times 10 \times 10 \qquad \text{Write the power as a product.}$$

$$= 5.34 \times \boxed{} \qquad \text{Multiply.}$$

$$= \boxed{} \qquad \text{Multiply.}$$

So, $5.34 \times 10^4 = 53{,}400$.

Method 2 Move the decimal point.

Moving the decimal point one place to the right is a result of multiplying a number by 10. So, when you multiply a number by 10^4, the decimal point moves four places to the right.

The exponent on the power of 10 indicates the number of places, in this case, _____. A positive exponent indicates the direction, to the _____.

$$5.34 \times 10^4 = 53{,}400$$

Check

Write 9.931×10^5 in standard form.

Show your work here

 Think About It!
What is standard form?

 Talk About It!
Compare the two methods used to write the number in standard form.

Go Online You can complete an Extra Example online.

Example 2 Write Numbers in Standard Form

Write 3.27×10^{-3} in standard form.

Method 1 Write the power as a product.

$3.27 \times 10^{-3} = 3.27 \times 10^{-1} \times 10^{-1} \times 10^{-1}$ Product of Powers Property

$\qquad = 3.27 \times \frac{1}{10} \times \frac{1}{10} \times \frac{1}{10}$ Definition of negative exponents

$\qquad = 3.27 \times \frac{1}{1,000}$ Multiply the fractions.

$\qquad = 3.27 \div 1,000$ Definition of reciprocal

$\qquad = \boxed{}$ Divide.

So, $3.27 \times 10^{-3} = 0.00327$.

Method 2 Move the decimal point.

When you divide a number by 10, the decimal point moves one place to the left. Dividing by 10 is the same as multiplying by 10^{-1}. So, when you multiply a number by 10^{-3}, the decimal point moves three places to the left.

The exponent on the power of 10 indicates the number of places, in this case, _____. A negative sign on the exponent indicates the direction, to the _____.

$3.27 \times 10^{-3} = 0.00327$

Check

Write 6.02×10^{-4} in standard form.

Show your work here

Go Online You can complete an Extra Example online.

Think About It!

How would you begin solving the problem?

Talk About It!

Compare the two methods used to write the number in standard form.

Learn Scientific Notation and Technology

Go Online Watch the video to learn how to interpret scientific notation that has been generated by a calculator.

The animation shows that a calculator will convert numbers to scientific notation only if they are very large or very small. The value after the E represents the exponent on the power of 10.

8E11 means $8 \times$ [   ]   2.5E−12 means $2.5 \times$ [   ]

You can also use a calculator to enter a number in scientific notation.

Step 1 Type the number that is multiplied by the power of 10.

Step 2 Press the 2ND button.

Step 3 Press the EE button, located above the comma button. (Note this will only display as E on the screen.)

Step 4 Type the exponent on the power of 10.

Think About It!

What do you think the symbol E represents on the calculator screen?

Example 3 Scientific Notation and Technology

Calculators often use the E symbol to indicate scientific notation.

The calculator screen shows the approximate wavelength, in meters, of violet light.

Write this number in standard form.

$$4 \text{E} -7$$

Step 1 Represent the number on the calculator screen in scientific notation.

4E−7 $\longrightarrow$ [ ]$\cdot$ $\times$ [ ]   4 is the factor and −7 is the exponent.

Step 2 Write the number in standard form.

To write 4×10^{-7} in standard form, move the decimal point _____ places to the _____.

So, 4×10^{-7} is 0.0000004.

Check

The calculator screen shows the diameter of a grain of sand in inches. Which numbers represent the number shown on the calculator screen? Select all that apply.

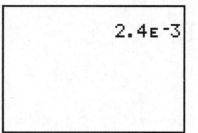

☐ 2,400

☐ 0.0024

☐ 2.4×10^3

☐ 2.4×10^{-3}

☐ 2.4^{-3}

 Go Online You can complete an Extra Example online.

Learn Write Numbers in Scientific Notation

When writing a positive number in scientific notation, the sign of the exponent can be determined by examining the number in standard form.

If the number in standard form is . . .

	greater than or equal to 10,	between 0 and 1,
Words	then the exponent in the power of 10 is *positive*.	then the exponent in the power of 10 is *negative*.
Example	$5,860,000 = 5.86 \times 10^6$	$0.000586 = 5.86 \times 10^{-4}$

💬 **Talk About It!**

If the number in standard form is greater than or equal to 1 and less than 10, what is the exponent on the power of 10 when the number is written in scientific notation? Explain.

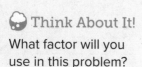

Example 4 Write Numbers in Scientific Notation

Write 3,725,000 in scientific notation.

$$3{,}725{,}000 = 3.725 \times 1{,}000{,}000 \qquad \text{Write the number as a product.}$$

$$= 3.725 \times \boxed{} \qquad \text{Since 3,725,000 is } >10\text{, the exponent is positive.}$$

So, 3,725,000 written in scientific notation is 3.725×10^6.

Check

Which expression represents 8,785,000,000 in scientific notation?

(A) 8.785×10^9 (C) 8.785×10^{-9}

(B) 8.785×10^6 (D) 87.85×10^8

Example 5 Write Numbers in Scientific Notation

Write 0.000316 in scientific notation.

$$0.000316 = 3.16 \times 0.0001 \qquad \text{Write the number as a product.}$$

$$= 3.16 \times \dfrac{1}{\boxed{}} \qquad \text{Write the decimal as a fraction with power of 10.}$$

$$= 3.16 \times \boxed{} \qquad \text{Since } 0 < 0.000316 < 1\text{, the exponent is negative.}$$

So, 0.000316 written in scientific notation is 3.16×10^{-4}.

Check

Which expression represents 0.524 in scientific notation?

(A) 5.24×10^1 (C) 5.24×10^{-1}

(B) 5.24×10^3 (D) 524×10^{-3}

Go Online You can complete an Extra Example online.

Think About It!
What factor will you use in this problem?

Talk About It!
Why was the decimal point moved 6 places? How can you determine the value of the exponent?

Think About It!
What factor will you use in this problem?

Talk About It!
Why was the decimal point moved 4 places? Why is the exponent negative?

Learn Use Scientific Notation

When using very large or very small quantities, it is important to choose the appropriate size of measurement.

For example, it is more appropriate to represent the distance from California to New York as 2,441 miles as opposed to 1.55×10^8 inches.

Circle the units that are more appropriate for measuring the following.

the time it takes to travel from Florida to Michigan

minutes   hours

the thickness of a penny

millimeters   meters

the length of a football field

inches   yards

the weight of a paperclip

ounces   pounds

You can estimate very large or very small numbers by expressing them in the form of a single digit times an integer power of 10. Estimating very large or very small numbers makes them easier to work with.

Complete the examples to estimate the numbers.

The population of the United States in a recent year was 324,430,860.

$324{,}430{,}860 \approx 300{,}000{,}000$

$300{,}000{,}000 = \boxed{} \times \boxed{}$

The diameter of an animal cell was measured at 0.000635 centimeter.

$0.000635 \approx 0.0006$

$0.0006 = \boxed{} \times \boxed{}$

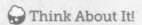

Think About It!

Which of the expressions is more meaningful to you?

Talk About It!

Describe some situations in which it might make sense to use a unit of measure that was *not* as meaningful.

 Example 6 Choose Units of Appropriate Size

If you could walk at a rate of 2 meters per second, it would take you 1.92×10^8 seconds to walk to the Moon.

Is it more appropriate to report this time as 1.92×10^8 seconds or 6.09 years? Explain your reasoning.

Representing the time in seconds gives a large number, 1.92×10^8 seconds or 192,000,000 seconds. It would be more meaningful to report the time as 6.09 years, so choosing the larger unit of measure is more appropriate in this situation.

So, the measure _____ is the more appropriate time.

Check

A plant cell has a diameter of 1.3×10^{-8} kilometer. Is it more appropriate to report the diameter of a plant cell as 1.3×10^{-8} kilometer or 1.3×10^{-2} millimeter?

Ⓐ 1.3×10^{-8} kilometer   Ⓑ 1.3×10^{-2} millimeter

Show your work here

 Example 7 Estimate with Scientific Notation

The population of Missouri is 6,063,589 people.

Write an estimation in scientific notation for the population.

$6,063,589 \approx$ [   ]   Estimate.

$6,000,000 =$ [   ] $\times$ [   ]   Write in scientific notation.

So, the population of Missouri is about 6,000,000 or 6×10^6 people.

Check

The distance from the Earth to the Moon is 238,900 miles. Write an estimation in scientific notation for the distance.

Show your work here

Go Online You can complete an Extra Example online.

⊕ Apply Travel

The table shows the approximate number of international visitors to the United States, from several countries, for July and August of a recent year. Which country had the greatest percent of change in the approximate number of visitors from July to August?

Country	Visitors (Jul)	Visitors (Aug)
Canada	2.09×10^6	2.43×10^6
Mexico	1.55×10^6	1.63×10^6
United Kingdom	4.2×10^5	4.9×10^5
Japan	3.05×10^5	3.9×10^5

1 What is the task?

Make sure you understand exactly what question to answer or problem to solve. You may want to read the problem three times. Discuss these questions with a partner.

First Time Describe the context of the problem, in your own words.
Second Time What mathematics do you see in the problem?
Third Time What are you wondering about?

2 How can you approach the task? What strategies can you use?

Record your observations here

3 What is your solution?

Use your strategy to solve the problem.

Show your work here

4 How can you show your solution is reasonable?

⊘ **Write About It!** Write an argument that can be used to defend your solution.

Talk About It! What method did you use to solve the problem? Explain why you chose that method.

Check

The approximate attendance and average ticket price for four Major League baseball teams for a recent year is shown in the table.

Team	Attendance	Ticket Price ($)
Los Angeles Angels	3,020,000	32.70
Miami Marlins	1.71×10^6	28.31
Pittsburgh Pirates	2,450,000	29.96
St. Louis Cardinals	3.44×10^6	34.20

Which team had the most revenue from ticket sales?

Show your work here

Go Online You can complete an Extra Example online.

Pause and Reflect

What have you learned about scientific notation in this lesson? Include a real-world example of scientific notation in your summary.

Record your observations here

Practice

Go Online You can complete your homework online.

Write each number in standard form. (Examples 1 and 2)

1. $1.6 \times 10^3 =$ _____

2. $1.49 \times 10^{-7} =$ _____

3. A calculator screen shows a number in scientific notation as 8.3E−6. Write this number in standard form. **(Example 3)**

4. A calculator screen shows a number in scientific notation as 7E11. Write this number in standard form. **(Example 3)**

Write each number in scientific notation. (Examples 4 and 5)

5. $2,204,000,000 =$ _____

6. $0.0000000642 =$ _____

7. A common race is a 5K race, where runners travel 5 kilometers. Is it more appropriate to report the distance as 5 kilometers or 5×10^6 millimeters? Explain your reasoning. **(Example 6)**

8. The population of Florida was recently recorded as 20,612,439 people. Write an estimation in scientific notation for the population. **(Example 7)**

Test Practice

9. The diameter of a grain of sand is 0.0024 inch. Write an estimation in scientific notation for the diameter. **(Example 7)**

10. Equation Editor The mass of planet Earth is about 5.98×10^{24} kilograms. When this number is written in standard notation, how many zeros are in the number?

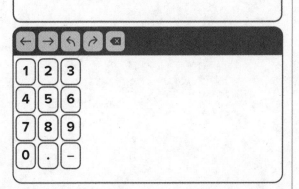

Apply

11. The table shows the approximate number of insects in each group. What is the combined population of the four insect groups?

Insect	Population
Honeybees	7.3498×10^4
Ants	6.822×10^3
Termites	9.8×10^5
Aphids	2.9502×10^4

12. The table shows the approximate population and land area for several states. Which state has the least population density?

State	Population	Land Area (mi^2)
Arizona	6.83×10^6	113,594
Colorado	5.46×10^6	103,642
Kentucky	4.43×10^6	39,486
Minnesota	5.45×10^6	79,627

13. Simplify the following expression and write in scientific notation.

$$\frac{(0.00045)(500,000)}{0.0015}$$

14. (MP) **Be Precise** The amount of time spent doing homework is 2.7×10^3 seconds. Choose a more appropriate measurement to report the amount of time spent doing homework. Justify your answer.

15. (MP) **Find the Error** Katrina states that 3.5×10^4 is greater than 2.1×10^6 because $3.5 > 2.1$. Explain her mistake and correct it.

16. Give a number in scientific notation that is between the two numbers on a number line.

$$7.1 \times 10^3 \text{ and } 71,000,000$$

Compute with Scientific Notation

I Can... perform computations with numbers written in scientific notation.

Learn Multiply and Divide with Scientific Notation

To solve real-world problems involving numbers written in scientific notation, you may need to multiply or divide these numbers. Since numbers written in scientific notation contain exponents, sometimes the Laws of Exponents can be used when operating with them.

With all operations, it is often helpful to write the numbers in the same form, scientific notation or standard form, before computing.

When performing these operations with numbers written in scientific notation...	Remember to...
Multiplication	Use the Product of Powers Property.
Division	Use the Quotient of Powers Property.

Example 1 Multiply with Scientific Notation

Scientists estimate that there are over 3.5×10^6 ants per acre in the Amazon rain forest.

If the Amazon rain forest covers approximately 1 billion acres, find the total number of ants. Write in scientific notation.

Step 1 Write the number of acres in scientific notation.

$$1 \text{ billion} = \boxed{} \times \boxed{}$$

Step 2 Multiply to find the total number of ants.

$(3.5 \times 10^6) \times (1 \times 10^9)$	Write the expression.
$= (3.5 \times 1) \times (10^6 \times 10^9)$	Commutative and Associative Properties
$= (3.5) \times (10^6 \times 10^9)$	Multiply 3.5 by 1.
$= 3.5 \times 10^{6 + 9}$	Product of Powers Property
$= 3.5 \times \boxed{}$	Add the exponents.

So, the total number of ants is approximately 3.5×10^{15}.

Think About It!

How will you set up the multiplication to solve this problem?

Talk About It!

In the 2nd step, why were the Commutative and Associative Properties used?

Check

A dime is 1.35×10^{-3} meter thick. What would be the height, in meters, of a stack of 1 million dimes?

Ⓐ 1.35×10^2 meters

Ⓑ 1.35×10^3 meters

Ⓒ 1.35×10^4 meters

Ⓓ 1.35×10^9 meters

Show your work here

🔵 **Go Online** You can complete an Extra Example online.

🌐 **Example 2** Divide with Scientific Notation

Neurons are cells in the nervous system that process and transmit information. An average neuron is about 5×10^{-6} meter in diameter. A standard table tennis ball is 0.04 meter in diameter.

About how many times as great is the diameter of a ball than a neuron? Write your answer in scientific notation.

Step 1 Write the numbers in the same form.

Write the diameter of the table tennis ball in scientific notation.

$0.04 = \boxed{} \times \boxed{}$

Step 2 Divide the diameter of the ball by the diameter of the neuron.

$\dfrac{4 \times 10^{-2}}{5 \times 10^{-6}} = \left(\dfrac{4}{5}\right) \times \left(\dfrac{10^{-2}}{10^{-6}}\right)$ Associative Property

$= \boxed{} \times \left(\dfrac{10^{-2}}{10^{-6}}\right)$ Divide 4 by 5.

$= 0.8 \times 10^{-2-(-6)}$ Quotient of Powers Property

$= 0.8 \times \boxed{}$ Subtract the exponents.

Step 3 Write in scientific notation.

Since 0.8×10^4 is not written in scientific notation, move the decimal one place to the right and subtract 1 from the exponent.

$0.8 \times 10^4 = \boxed{} \times \boxed{}$

So, the diameter of the table tennis ball is about 8×10^3 or _____ times larger than the diameter of the neuron.

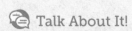

😮 **Think About It!**

How would you begin solving the problem?

💬 **Talk About It!**

Why is 0.8×10^4 not in scientific notation? When writing 0.8×10^4 in scientific notation, why do you need to subtract 1 from the exponent?

Check

In a recent year, the population of China was about 1.3×10^9. According to the census data, the population of the United States was about 308,745,538. About how many times greater was the population of China than the population of the United States?

Show your work here

 Go Online You can complete an Extra Example online.

Learn Add and Subtract with Scientific Notation

To add or subtract numbers written in scientific notation, it is helpful to rewrite the numbers so that the exponents on each power of 10 have the same value.

To add 3.6×10^4 to the numbers shown, each number needs to be rewritten so the exponents on each power of 10 are the same. Rewrite each number so that it has an exponent of 4.

$$1.43 \times 10^3 = \boxed{} \times 10^4$$

$$1.43 \times 10^5 = \boxed{} \times 10^4$$

$$1.43 \times 10^2 = \boxed{} \times 10^4$$

Once numbers written in scientific notation have the same exponent on the power of 10, you can use the Distributive Property to add or subtract.

$(4.36 \times 10^5) + (2.09 \times 10^6)$

$\quad = (4.36 \times 10^5) + (20.9 \times 10^5)$ Rewrite with the same power of 10.

$\quad = (4.36 + 20.9) \times 10^5$ Distributive Property

$\quad = (25.26) \times 10^5$ Add 4.36 and 20.9.

$\quad = 2.526 \times 10^6$ Rewrite in scientific notation.

You can also write each number in standard form before computing or to check your work.

Talk About It!

Explain why you cannot add the exponents on the powers of 10, when adding the expressions.

Think About It!

Do the powers of 10 have the same exponent?

Talk About It!

Describe another method you can use to evaluate the expression.

Example 3 Add or Subtract with Scientific Notation

Evaluate $(1.45 \times 10^9) - (2.84 \times 10^8)$. Express the result in scientific notation.

$(1.45 \times 10^9) - (2.84 \times 10^8)$ Write the expression.

$= (14.5 \times 10^8) - (2.84 \times 10^8)$ Write 1.45×10^9 as 14.5×10^8.

$= (\boxed{} - \boxed{}) \times 10^8$ Distributive Property

$= \boxed{} \times 10^8$ Subtract 2.84 from 14.5.

$= \boxed{} \times 10^9$ Rewrite in scientific notation.

So, $(1.45 \times 10^9) - (2.84 \times 10^8) = 1.166 \times 10^9$.

Check

Which of the following is the sum $(8.41 \times 10^3) + (9.71 \times 10^4)$ written in scientific notation?

(A) 1.0551×10^5

(B) 1.0551×10^7

(C) 10.551×10^4

(D) 105.51×10^3

 Go Online You can complete an Extra Example online.

Pause and Reflect

How is addition of numbers written in scientific notation similar to subtraction of numbers written in scientific notation? How is it different?

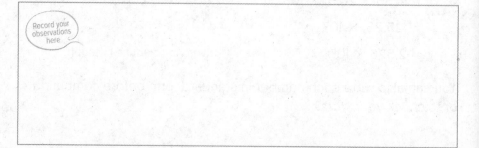

Apply Population

In a recent year, the world population was about 7,289,332,000. The population of the United States was about 3×10^8. About how many times larger was the world population than the population of the United States?

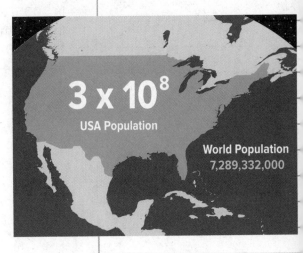

3 x 10⁸
USA Population

World Population
7,289,332,000

1 What is the task?

Make sure you understand exactly what question to answer or problem to solve. You may want to read the problem three times. Discuss these questions with a partner.

First Time Describe the context of the problem, in your own words.
Second Time What mathematics do you see in the problem?
Third Time What are you wondering about?

2 How can you approach the task? What strategies can you use?

Record your observations here

3 What is your solution?

Use your strategy to solve the problem.

Show your work here

🗨 Talk About It!

Why might it be helpful to round the world population?

4 How can you show your solution is reasonable?

⬦ **Write About It!** Write an argument that can be used to defend your solution.

Check

In 2005, 8.1×10^{10} text messages were sent in the United States. In 2010, the number of annual text messages had risen to 1,810,000,000,000. About how many times as great was the number of text messages in 2010 than 2005?

Show your work here

🔾 **Go Online** You can complete an Extra Example online.

Pause and Reflect

Have you ever wondered when you might use the concepts you learn in math class? What are some everyday scenarios in which you might use what you learned today?

Record your observations here

Practice

Go Online You can complete your homework online.

1. There are about 3×10^{11} stars in our galaxy and about 100 billion galaxies in the observable universe. Suppose every galaxy has as many stars as ours. How many stars are in the observable universe? Write in scientific notation. **(Example 1)**

2. Humpback whales are known to weigh as much as 80,000 pounds. The tiny krill they eat weigh only 2.1875×10^{-3} pound. About how many times greater is the weight of a humpback whale? **(Example 2)**

Evaluate. Express each result in scientific notation. (Example 3)

3. $(1.28 \times 10^5) + (1.13 \times 10^3) =$

_____

4. $(7.26 \times 10^6) - (1.3 \times 10^4) =$

_____

Test Practice

5. The speed of light is about 1.86×10^5 miles per second. The star Sirius is about 5.062×10^{13} miles from Earth. About how many seconds does it take light to travel from Sirius to Earth? Write in scientific notation, rounded to the nearest hundredth.

6. **Table Item** The table shows the amount of money raised by each region. The four regions raised a total of (5.38×10^4). How much did the West raise?

Region	Amount Raised ($)
East	1.46×10^4
North	2.38×10^4
South	6.75×10^3
West	

Apply

7. A bacterium was found to have a mass of 2×10^{-12} gram. After 30 hours, one bacterium was replaced by a population of 480,000,000 bacteria. What is the mass of the population of bacteria after 30 hours? Write your answer in scientific notation.

8. At the beginning of the business day, a bank's vault held $575,900. By the end of the day, (3.5×10^3) had been added to the vault. How much money did the bank's vault hold at the end of the business day? Write your answer in standard form.

9. Explain how the Product of Powers and Quotient of Powers Properties help you to multiply and divide numbers in scientific notation.

10. MP Persevere with Problems
An Olympic-sized swimming pool holds 6.6043×10^5 gallons of water. A standard garden hose can deliver 9 gallons of water per minute. If the garden hose was filling the pool 24 hours per day, how many days would it take to fill the Olympic-sized pool? Explain your reasoning.

11. MP Find the Error Miguel found the quotient $\frac{5.78 \times 10^5}{0.000002}$ as 0.289. Find his mistake and correct it.

12. MP Be Precise A *googol* is the number 1 followed by 100 zeros. Earth's mass is 5.972×10^{24} kilograms. How many Earths are needed to have a total mass of 1 googol kilograms?

Review

📖 Foldables Use your Foldable to help review the module.

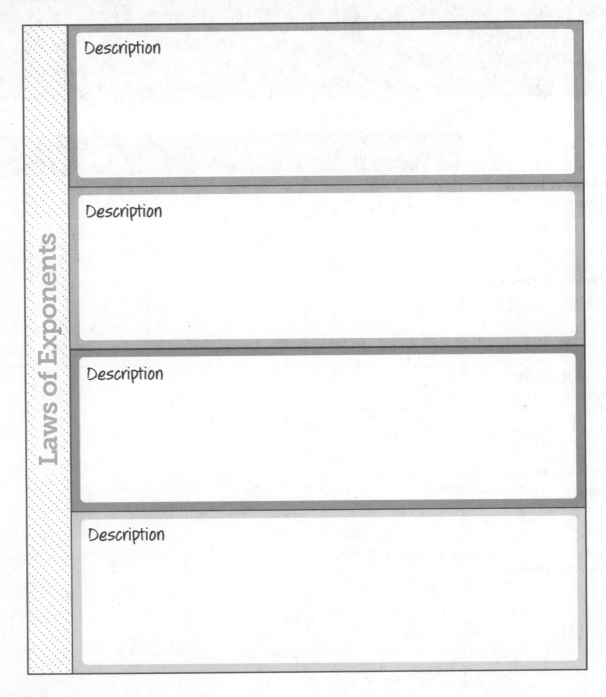

Laws of Exponents

Description

Description

Description

Description

Rate Yourself! ⬛ ◆ ★

Complete the chart at the beginning of the module by placing a checkmark in each row that corresponds with how much you know about each topic after completing this module.

Reflect on the Module

Use what you learned about exponents and scientific notation to complete the graphic organizer.

e Essential Question

Why are exponents useful when working with very large or very small numbers?

	Words	Algebra	Numbers
Product of Powers Property			
Quotient of Powers Property			
Power of a Power Property			
Power of a Product Property			
Scientific Notation			

Name _____ Period _____ Date _____

Test Practice

1. Multiple Choice Which of the following represents the expression
$a \cdot b \cdot a \cdot a \cdot b \cdot a \cdot b \cdot b \cdot a \cdot b \cdot b \cdot b$ using exponents? (Lesson 1)

(A) $a^7 \cdot b^5$

(B) $a^5 \cdot b^7$

(C) $a^5 \cdot b^5$

(D) $a^7 \cdot b^7$

2. Equation Editor The table shows the number of admission tickets to an amusement park sold over the last 3 days. How many more tickets were sold on Sunday than on Friday? (Lesson 1)

Day	Tickets Sold
Friday	$2^4 \cdot 3^2 \cdot 5^3$
Saturday	$2^2 \cdot 3^4 \cdot 7 \cdot 13$
Sunday	$3^6 \cdot 5 \cdot 7$

3. Open Response Louis is comparing the populations of several nearby cities. The population of Liberty Crossing can be written as 4^9 people, and the population of Harrisburg can be written as 4^7 people. The population of Glenview is 4^3 times the population of Harrisburg. Write the names of the cities from the city with the least population to the city with the greatest population. (Lesson 2)

4. Open Response Simplify the expression $(-3m^2)(-5m^7)$. Explain your process and justify each step by naming the appropriate property. (Lesson 2)

5. Table Item Indicate whether each statement is *true* or *false*. (Lesson 3)

	true	false
$(d^4)^3 = d^{12}$		
$(n^5)^{10} = n^{15}$		
$(k^7)^3 = k^{21}$		

6. Multiselect A square room has side lengths that can be represented by the expression $6x^3$ feet. Kyra wants to cover the floor in square tiles with side lengths that can be represented by the expression $3x$ feet. (Lesson 3)

A. Which of the following statements are correct? Select all that apply.

☐ The area of one tile is $3x^2$ ft².

☐ The area of one tile is $9x^2$ ft².

☐ The area of the floor is $36x^3$ ft².

☐ The area of the floor is $18x^4$ ft².

☐ The area of the floor is $36x^6$ ft².

B. If $x = 3$, then what number of tiles will be needed to cover the floor?

7. Open Response Express the fraction $\frac{1}{6^4}$ using a negative exponent. (Lesson 4)

8. Table Item Place an X in the correct column to indicate if each statement is *sometimes*, *always*, or *never* true. (Lesson 4)

	sometimes	always	never
$4^0 = 1$			
$n^0 = 1$			
$z^0 = 1, z \neq 0$			

9. Multiple Choice Which expression represents $w^{-7} \cdot w^3$ written in simplest form? (Lesson 4)

Ⓐ $\frac{1}{w^4}$ Ⓒ $\frac{1}{w^{-4}}$

Ⓑ w^{-4} Ⓓ w^4

10. Equation Editor Write 33,800,000 in scientific notation. (Lesson 5)

11. Open Response Write 7.2×10^{-3} in standard form. (Lesson 5)

12. Multiple Choice Which of the following is most appropriate to describe the average distance from Earth to the Sun? (Lesson 5)

Ⓐ 1.5×10^{14} mm

Ⓑ 1.5×10^{13} cm

Ⓒ 1.5×10^{11} m

Ⓓ 1.5×10^{8} km

13. Open Response Evaluate the expression $(7.2 \times 10^7) + (5.5 \times 10^6)$. Express the result in scientific notation. (Lesson 6)

14. Equation Editor Ryan downloaded a video game to his computer. The size of the file was 2.4×10^5 kilobytes. After the game was installed, he downloaded a patch file for the game that was 48,000 kilobytes in size. (Lesson 6)

A. What is the size of the game in standard form?

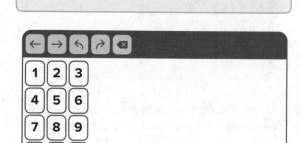

B. How many times larger was the game than the patch file?

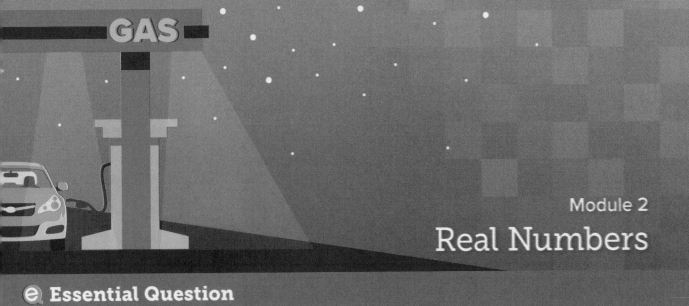

Real Numbers

e Essential Question

Why do we classify numbers?

What Will You Learn?

Place a checkmark (✓) in each row that corresponds with how much you already know about each topic **before** starting this module.

KEY	Before			After		
▢ — I don't know. ◊ — I've heard of it. ★ — I know it!	▢	◊	★	▢	◊	★
writing fractions and mixed numbers as decimals						
writing decimals as fractions and mixed numbers						
finding square roots and cube roots						
identifying real numbers						
describing sets of real numbers						
estimating irrational numbers						
comparing and ordering real numbers						
graphing real numbers on a number line						

▥ Foldables Cut out the Foldable and tape it to the Module Review at the end of the module. You can use the Foldable throughout the module as you learn about real numbers.

What Vocabulary Will You Learn?

Check the box next to each vocabulary term that you may already know.

☐ bar notation

☐ counterexample

☐ cube root

☐ integers

☐ inverse operations

☐ irrational number

☐ natural numbers

☐ perfect cube

☐ perfect square

☐ principal square root

☐ radical sign

☐ rational numbers

☐ real number

☐ repeating decimal

☐ square root

☐ terminating decimal

☐ truncating

☐ whole numbers

Are You Ready?

Study the Quick Review to see if you are ready to start this module.
Then complete the Quick Check.

Quick Review	
Example 1 **Classify numbers.** Which numbers in the following list are natural numbers? $$5, -7, \frac{1}{2}, -\frac{2}{3}, 0, 2$$ A natural number is a counting number, such as 1, 2, 3, ... So, 5 and 2 are natural numbers.	**Example 2** **Compare rational numbers.** Fill in the blank with <, >, or = to make $\frac{3}{5}$ _____ 0.6666... a true statement. Since $\frac{3}{5}$ = 0.6, and 0.6 is less than 0.6666..., $\frac{3}{5}$ < 0.6666...

Quick Check	
1. The temperature fell 6°F, which can be represented by −6. Is −6 a natural number or integer?	**2.** Fill in the blank with <, >, or = to make $-2\frac{1}{3}$ _____ −2.5 a true statement.

How Did You Do?
Which exercises did you answer correctly in the Quick Check?
Shade those exercise numbers at the right.

Terminating and Repeating Decimals

I Can... show how the decimal form of a rational number repeats eventually, and convert a repeating decimal into a rational number.

What Vocabulary Will You Learn?

bar notation

integers

natural numbers

rational numbers

repeating decimal

terminating decimal

whole numbers

Explore Terminating Decimals

Online Activity You will explore what it means for a decimal to be a terminating decimal.

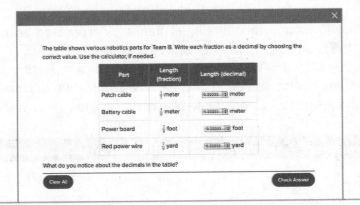

Learn Rational Numbers

When you first learned to count using the numbers 1, 2, 3, ..., you were using members of the set of **natural numbers**. If you add zero to the set of natural numbers, the result is the set of **whole numbers**, {0 , 1 , 2 , 3 , . . . }. Whole numbers and their opposites make up the set of **integers**, { . . . , −3, −2, −1, 0 , 1 , 2 , 3 , . . . }. Natural numbers, whole numbers, and integers are part of the set of **rational numbers**.

Words	Model
A rational number is a number that can be written as the ratio of two integers in which the denominator is not zero.	Rational Numbers
Symbols	
$\frac{a}{b}$, where a and b are integers and $b \neq 0$.	

Learn Terminating and Repeating Decimals

Every rational number can be expressed as a decimal by dividing the numerator by the denominator. The decimal form of a rational number either terminates in 0s or eventually repeats. **Repeating decimals** are decimals in which 1 or more digits repeat. If the repeating digit is zero, then the decimal is a **terminating decimal**.

Bar notation is often used to indicate that a digit or group of digits repeats. The bar is placed above the digit(s) that repeat. For example, 0.833333333... can be written as $0.8\overline{3}$.

Go Online Watch the animation to see examples of terminating and repeating decimals.

The animation shows that when a rational number is expressed as a decimal, and the repeating digit is not zero, the decimal does not terminate. To complete the table, write each fraction as a repeating decimal. Then write each repeating decimal as a terminating decimal. If the decimal does not terminate, write *does not terminate*.

Rational Number	Repeating Decimal	Terminating Decimal
$\frac{2}{5}$		
$\frac{5}{6}$		

Write each decimal from the list below in the correct box.

0.64000..., $0.\overline{56}$,   0.4242...,

0.84,   0.333...,   0.7

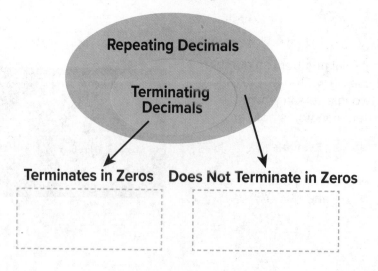

Terminates in Zeros   Does Not Terminate in Zeros

Example 1 Write Fractions as Decimals

Write $\frac{5}{8}$ as a decimal. Then determine whether the decimal is a terminating decimal.

Part A Write the fraction as a decimal.

$\frac{5}{8}$ means 5 ÷ 8. Divide 5 by 8.

```
    0.625
8) 5.000
  −4 8
     20
    −16
     40
    −40
      0
```

Place the decimal point. Annex zeros and divide as with whole numbers.

So, $\frac{5}{8}$ = _____.

Part B Determine whether the decimal is a terminating decimal.

The repeating digit is _____. If you were to continue dividing, the quotient would be 0.625000... .

So, 0.625 is a terminating decimal.

Check

Write $-\frac{2}{9}$ as a decimal. Then determine whether the decimal is a terminating decimal.

Part A

Which decimal is equivalent to $-\frac{2}{9}$?

Ⓐ −0.2

Ⓑ −0.22

Ⓒ −0.$\overline{2}$

Ⓓ −0.23

Part B

Determine whether the decimal is a terminating decimal.

(Show your work here)

 Go Online You can complete an Extra Example online.

Think About It!

How would you begin writing the fraction as a decimal?

Talk About It!

Why is the decimal form of $\frac{5}{8}$ also considered a repeating decimal?

Think About It!
How would you begin writing the mixed number as a decimal?

Talk About It!
When using long division to write $-1\frac{2}{3}$ as a decimal, how do you know when you can stop dividing?

Example 2 Write Mixed Numbers as Decimals

Write $-1\frac{2}{3}$ as a decimal. Then determine whether the decimal is a terminating decimal.

Part A Write $-1\frac{2}{3}$ as a decimal.

$-1\frac{2}{3}$ can be rewritten as $\dfrac{\square}{\square}$. Divide 5 by 3 and add a negative sign.

$$
\begin{array}{r}
1.6... \\
3\overline{)5.0} \\
-3 \\
\hline
20 \\
-18 \\
\hline
2
\end{array}
$$

The remainder will never be 0.

So, $-1\frac{2}{3} = -1.\overline{6}$.

Part B Determine whether the decimal is a terminating decimal.

The repeating digit is _____.

So, $-1.\overline{6}$ is *not* a terminating decimal because the repeating digit is not zero.

Check

Write $4\frac{13}{25}$ as a decimal. Then determine whether the decimal is a terminating decimal.

Part A

Which decimal is equivalent to $4\frac{13}{25}$?

Ⓐ 4.5

Ⓑ 4.52

Ⓒ $4.\overline{52}$

Ⓓ 4.525

Part B

Determine whether the decimal is a terminating decimal.

Show your work here

🅝 **Go Online** You can complete an Extra Example online.

Learn Write Repeating Decimals as Fractions

 Go Online Watch the animation to learn how to use an algebraic method to write a repeating, non-terminating decimal, such as $0.\overline{4}$, as a fraction.

$N = 0.444...$	Assign a variable to the value of the decimal.
$10(N) = 10(0.444...)$	Multiply each side by a power of 10.
$10N = 4.444...$	Multiplying by 10 moves the decimal point one place to the right.
$-(N = 0.444...)$	Subtract the original equation to eliminate the repeating part.
$9N = 4$	Simplify.
$\dfrac{9N}{9} = \dfrac{4}{9}$	Divide each side by 9.
$N = \dfrac{4}{9}$	Simplify.

The decimal 0.444... is equivalent to $\dfrac{4}{9}$.

Talk About It!

How do you know that $0.\overline{4}$ cannot be written as the fraction $\dfrac{44}{100}$ nor the fraction $\dfrac{444}{1,000}$?

Example 3 Write Repeating Decimals as Fractions

Write $0.\overline{5}$ as a fraction in simplest form.

Assign a variable to the value $0.\overline{5}$. Let $N = 0.555...$. Then perform operations on N to determine its fractional value.

$N = 0.555...$	
$10(N) = 10(0.555...)$	Multiply each side by 10 because one digit repeats.
$10N = 5.555...$	Multiplying by 10 moved the decimal point one place to the right.
$-(N = 0.555...)$	Subtract $N = 0.555...$ to eliminate the repeating part.
$9N = 5$	Simplify.
$N = \dfrac{\boxed{}}{\boxed{}}$	Divide each side by 9.

So, the decimal $0.\overline{5}$ can be written as the fraction $\dfrac{5}{9}$.

Check

Write $-0.\overline{7}$ as a fraction in simplest form.

Talk About It!

How can you verify that you determined the correct fractional value?

 Go Online

You can complete an Extra Example online.

Example 4 Write Repeating Decimals as Mixed Numbers

Write 2.$\overline{18}$ as a mixed number in simplest form.

Assign a variable to the value 2.$\overline{18}$. Let N = 2.181818... . Then perform operations on N to determine its fractional value.

$$N = 2.181818...$$

$$100(N) = 100(2.181818...)$$ Multiply each side by 100 because two digits repeat.

$$100N = 218.181818...$$ Simplify.

$$-(N = 2.181818...)$$ Subtract N = 2.181818... to eliminate the repeating part.

$$99N = \boxed{}$$ Simplify.

$$N = \frac{\boxed{}}{\boxed{}}$$ Divide each side by 99.

$$N = \boxed{}$$ Write as a mixed number in simplest form.

So, the decimal 2.$\overline{18}$ can be written as $2\frac{2}{11}$.

Check

Write 1.$\overline{42}$ as a mixed number in simplest form.

Talk About It!

In the second line of the solution, why do you multiply by 100 instead of 10?

Go Online You can complete an Extra Example online.

🌐 **Apply** Baseball

In a recent season, a first baseman had 175 hits in 530 at-bats. At this rate, how many hits would he have in 590 at-bats?

Player Stats ⚾	
At Bats (AB)	Hits (H)
530	175
590	?

1 What is the task?

Make sure you understand exactly what question to answer or problem to solve. You may want to read the problem three times. Discuss these questions with a partner.

First Time Describe the context of the problem, in your own words.
Second Time What mathematics do you see in the problem?
Third Time What are you wondering about?

2 How can you approach the task? What strategies can you use?

3 What is your solution?

Use your strategy to solve the problem.

4 How can you show your solution is reasonable?

✏️ **Write About It!** Write an argument that can be used to defend your solution.

💬 **Talk About It!**
How can you solve the problem another way?

Check

At hockey practice, Ajay saved 73 out of the 126 shots on goal. At this rate, how many saves would he make out of 200 shots on goal?

Show your work here

🡒 **Go Online** You can complete an Extra Example online.

📖 **Foldables** It's time to update your Foldable, located in the Module Review, based on what you learned in this lesson. If you haven't already assembled your Foldable, you can find the instructions on page FL1.

Real Numbers		
Rational	Irrational	

Write About It   Write About It

Practice

Go Online You can complete your homework online.

Write each rational number in decimal form. Then determine whether the decimal is a terminating decimal. (Examples 1 and 2)

1. $-\dfrac{11}{16} =$ _____

2. $\dfrac{5}{33} =$ _____

3. $4\dfrac{3}{8} =$ _____

4. $-9\dfrac{11}{30} =$ _____

Write each decimal as a fraction or mixed number in simplest form. (Examples 3 and 4)

5. $0.\overline{8} =$ _____

6. $-0.\overline{18} =$ _____

7. $-1.\overline{5} =$ _____

8. $4.\overline{45} =$ _____

9. One pint is about $\dfrac{5}{9}$ liter. Write $\dfrac{5}{9}$ as a decimal.

10. Phoebe won $\dfrac{7}{16}$ of the competitions she entered. Write $\dfrac{7}{16}$ as a decimal.

Test Practice

11. Write the decimal 5.666... as a mixed number in simplest form.

12. **Multiselect** Select all of the fractions that, when converted, result in a non-terminating decimal.

☐ $-\dfrac{11}{5}$ ☐ $\dfrac{1}{50}$

☐ $\dfrac{1}{225}$ ☐ $\dfrac{7}{26}$

☐ $\dfrac{5}{16}$ ☐ $-\dfrac{3}{32}$

Apply

13. In one soccer season, Milica saved 155 out of 180 shots on goal. At this rate, about how many saves would she make out of 600 shots on goal?

Saves	Shots on Goal
155	180
?	600

14. In a single tennis season, Akeem had 100 aces in 53 matches. At this rate, about how many aces would he have in 200 matches?

15. (MP) **Identify Structure** Give an example of a decimal where three digits repeat. Explain why it is a rational number.

16. (MP) **Use a Counterexample** If all integers are rational numbers, are all rational numbers integers? If yes, explain why. If not, give a counterexample.

17. Make a prediction about whether the rational number $\frac{3}{162}$ is a terminating or non-terminating decimal when written in decimal form. Explain your reasoning for your prediction. Then write the fraction as a decimal to check your prediction.

18. (MP) **Be Precise** You are asked *what is 49 divided by 11?* Write the answer in the most precise way. Explain why you chose this way to write the answer.

Roots

I Can... find square and cube roots, and use square and cube roots to solve equations involving perfect squares and cubes.

What Vocabulary Will You Learn?
cube root

inverse operations

perfect cube

perfect square

principal square root

radical sign

square root

Explore Find Square Roots Using a Square Model

Online Activity You will use Web Sketchpad to explore how to use square models to find square roots.

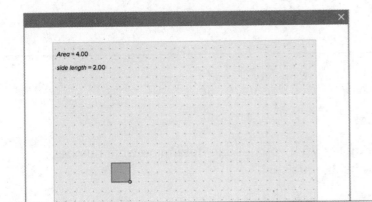

Area = 4.00

side length = 2.00

Learn Square Roots

Words
A **square root** of a number is one of its two equal factors.
Symbols
If $x^2 = y$, then x is the square root of y.
Example
$5^2 = 25$, so 5 is a square root of 25.

A **perfect square** is a rational number whose square root is a whole number. Complete the table for the following perfect squares.

Perfect Square	1	4	9	16	25	36	49	64
Square Root	1	2	3	4				

Talk About It!

Why do you think the term *perfect square* uses the adjective *perfect*? What are some numbers that aren't perfect squares? Explain why.

(continued on next page)

Every positive number has *both* a positive and negative square root.

Since 3 · 3 = 9, _____ is a square root of 9.

Since (–3) · (–3) = 9, _____ is a square root of 9.

Therefore, 9 has two square roots, 3 and –3.

In most real-world situations, only the positive or **principal square root** is considered. A **radical sign**, $\sqrt{\ }$, is used to indicate the principal square root. When both the positive and negative square roots are asked for, the ± symbol is used before the radical sign.

$$\sqrt{25} = \boxed{} \qquad -\sqrt{25} = \boxed{} \qquad \pm\sqrt{25} = \boxed{}$$

Example 1 Find Positive Square Roots

Simplify $\sqrt{64}$.

In order to simplify $\sqrt{64}$, you need to determine what number, multiplied by itself equals 64.

Find the factors of 64. _____

Find the square root.

$$\sqrt{64} = \boxed{} \qquad \text{Find the positive square root of 64; } 8^2 = 64.$$

So, $\sqrt{64} = 8$.

Check

Simplify $\sqrt{225}$.

Go Online You can complete an Extra Example online.

Example 2 Find Both Square Roots

Simplify $\pm\sqrt{1.21}$.

Step 1 Since there is an even number of decimal places, consider the square root of 121. Determine what number, multiplied by itself, equals 121.

$$\sqrt{121} = \boxed{}$$

Step 2 Determine where to place the decimal point in 11. There are only two options for placing the decimal point in the number 11, either before each digit (0.11) or between the digits (1.1). Since 1.21 has two decimal places, then the sum of the number of decimal places in each factor must equal 2.

If you multiply 0.11 · 0.11, then the product will have _____ decimal places.

If you multiply 1.1 · 1.1, then the product will have _____ decimal places.

So, $\pm\sqrt{1.21} = \pm 1.1$.

Check

Simplify $\pm\sqrt{1.44}$.

 Go Online You can complete an Extra Example online.

Example 3 Find Negative Square Roots

Simplify $-\sqrt{\dfrac{25}{36}}$.

Find the square root of $\dfrac{25}{36}$. Then add a negative sign.

$$-\sqrt{\dfrac{25}{36}} = -\dfrac{\boxed{}}{\boxed{}} \qquad \text{Find the negative square root of } \dfrac{25}{36};$$
$$\left(\dfrac{5}{6}\right)^2 = \dfrac{25}{36}.$$

So, $-\sqrt{\dfrac{25}{36}} = -\dfrac{5}{6}$.

Think About It!

What does the symbol $\pm$ before the radical sign indicate?

Check

Simplify $-\sqrt{\dfrac{49}{64}}$.

Think About It!

Can you think of a number when multiplied by itself, equals −16? Why or why not?

Example 4 Square Roots of Negative Numbers

Simplify $\sqrt{-16}$ using rational numbers. If the expression cannot be simplified, explain why.

In order to simplify $\sqrt{-16}$, you need to determine what number, multiplied by itself equals −16.

There is no rational number square root of −16 because

_____ times itself is equal to −16.

So, $\sqrt{-16}$ cannot be simplified and has no rational number solution.

Talk About It!

What is the difference between $\sqrt{-16}$ and $-\sqrt{16}$?

Check

Simplify $\sqrt{-81}$ using rational numbers. If the expression cannot be simplified, explain why.

Go Online You can complete an Extra Example online.

Learn Use Square Roots to Solve Equations

You can solve equations by using inverse operations. **Inverse operations** undo each other. Squaring and taking a square root are inverse operations.

Square a Number

$$9^2 = \boxed{}$$

Take the Square Root

$$\sqrt{81} = \boxed{}$$

To solve an equation of the form $x^2 = p$ for x, undo the operations of squaring x by taking the square root of each side.

$x^2 = p$	Write the equation.
$\pm\sqrt{x^2} = \pm\sqrt{p}$	Take the square root of each side.
$x = \pm p$	Simplify.

There will be two solutions, a positive square root and a negative square root.

Example 5 Use Square Roots to Solve Equations

Solve $t^2 = 169$. Check your solution.

$t^2 = 169$	Write the equation.
$\pm\sqrt{t^2} = \pm\sqrt{169}$	Take the square root of each side.
$t = \pm\boxed{}$	Definition of square root
$t = 13 \text{ or } -13$	Simplify.

So, the solutions to the equation are _____ and _____.

Check

Solve $y^2 = 256$.

Show your work here

Go Online You can complete an Extra Example online.

Talk About It!
Consider the equation $x^2 = 121$.

- If x is a positive number, what is x^2?
- If x is a negative number, what is x^2?
- How many solutions does the equation have? What are they?

Talk About It!
Why do we take the square root of each side of the equation? Why does this equation have two solutions?

Learn Cube Roots

Talk About It!

Why do you think the term *cube root* has the term *cube* in it?

Words
A **cube root** of a number is one of its three equal factors.
Symbols
If $x^3 = y$, then $x = \sqrt[3]{y}$.
Numbers
Since $2^3 = 8$, 2 is the cube root of 8.
Since $(-6)^3 = -216$, -6 is the cube root of -216.

Every integer has exactly one cube root. Complete the table that demonstrates this concept.

	Rule	Example
Cube Root of a Positive Number	The cube root of a positive number is positive.	$\sqrt[3]{27} = \boxed{}$ $\sqrt[3]{125} = \boxed{}$
Cube Root of Zero	The cube root of zero is zero.	$\sqrt[3]{0} = \boxed{}$
Cube Root of a Negative Number	The cube root of a negative number is negative.	$\sqrt[3]{-27} = \boxed{}$ $\sqrt[3]{-125} = \boxed{}$

A **perfect cube** is a number that is the cube of an integer.

Complete the table for the following perfect cubes.

Perfect Cube	1	−1	8	−8	27	−27	64	−64
Cube Root	1	−1	2	−2				

Pause and Reflect

Compare and contrast square roots and cube roots.

> Record your observations here

Example 6 Cube Roots of Positive Numbers

Simplify $\sqrt[3]{125}$.

In order to simplify $\sqrt[3]{125}$, you need to determine what number, multiplied 3 times, is equal to 125.

$\sqrt[3]{125} = \boxed{}$ $5^3 = 5 \cdot 5 \cdot 5$ or 125

So, $\sqrt[3]{125} = 5$.

Check

Simplify $\sqrt[3]{216}$.

Example 7 Cube Roots of Negative Numbers

Simplify $\sqrt[3]{-27}$.

In order to simplify $\sqrt[3]{-27}$ you need to determine what number, multiplied 3 times, is equal to −27.

$\sqrt[3]{-27} = \boxed{}$ $(-3)^3 = (-3) \cdot (-3) \cdot (-3)$ or −27

So, $\sqrt[3]{-27} = -3$.

Check

Simplify $\sqrt[3]{-1,000}$.

 Think About It!

Will the answer be a positive or negative number?

 Talk About It!

What is the difference between the cube root of a negative number and the square root of a negative number?

Go Online You can complete an Extra Example online.

Example 8 Use Cube Roots to Solve Equations

Dylan has a planter in the shape of a cube that holds 15.625, or $\frac{125}{8}$, cubic feet of potting soil.

Solve the equation $s^3 = \frac{125}{8}$ to find the side length s of the container. Check your solution.

To solve an equation of the form $x^3 = p$, take the cube root of each side of the equation.

$$s^3 = \frac{125}{8}$$ Write the equation.

$$\sqrt[3]{s^3} = \sqrt[3]{\frac{125}{8}}$$ Take the cube root of each side.

$$s = \frac{5}{2} \text{ or } \boxed{}$$ Definition of cube root

So, each side of the container is $2\frac{1}{2}$ feet.

Check the solution.

$$s^3 = \frac{125}{8}$$ Write the equation.

$$\left(\frac{5}{2}\right)^3 \overset{?}{=} \frac{125}{8}$$ Replace s with $\frac{5}{2}$.

$$\frac{125}{8} = \frac{125}{8}$$ Simplify. The solution, $\frac{5}{2}$, or $2\frac{1}{2}$, is correct.

Check

A box that is shaped like a cube has a volume of $\frac{512}{27}$ cubic inches. Solve $s^3 = \frac{512}{27}$ to find the length s of one side of the box.

Show your work here

Go Online You can complete an Extra Example online.

🌐 **Apply** Bulletin Boards

A bulletin board consists of four equal-sized cork squares arranged in a row to form a rectangle. If the total area of all four cork squares is 36 square feet, what is the length in feet of the bulletin board?

1 What is the task?

Make sure you understand exactly what question to answer or problem to solve. You may want to read the problem three times. Discuss these questions with a partner.

First Time Describe the context of the problem, in your own words.
Second Time What mathematics do you see in the problem?
Third Time What are you wondering about?

2 How can you approach the task? What strategies can you use?

Record your observations here

3 What is your solution?

Use your strategy to solve the problem.

Show your work here

4 How can you show your solution is reasonable?

✏️ **Write About It!** Write an argument that can be used to defend your solution.

Talk About It!

How would the length and width of the bulletin board change if the four cork squares were arranged in a square? How would the area be affected?

Check

A set of windows consists of three equal-sized squares arranged in a row to form a rectangle. If the total area of all three windows is 108 square feet, what is the length, in feet, of the windows?

Show your work here

Go Online You can complete an Extra Example online.

Pause and Reflect

Create a graphic organizer that will help you study the vocabulary and concepts from this lesson.

Record your observations here

Practice

Go Online You can complete your homework online.

Simplify using rational numbers. If the expression cannot be simplified, explain why. (Examples 1–4)

1. $\sqrt{361} =$ _____

2. $\pm\sqrt{1.96} =$ _____

3. $-\sqrt{\dfrac{9}{16}} =$ _____

4. $\sqrt{-441} =$ _____

5. Solve $m^2 = 0.04$. (Example 5) _____

Simplify using rational numbers. (Examples 6 and 7)

6. $\sqrt[3]{343} =$ _____

7. $\sqrt[3]{-512} =$ _____

8. A basin of a water fountain is cube shaped and has a volume of 91.125 cubic feet. Solve $s^3 = 91.125$ to find the length s of one side of the basin. (Example 8)

Test Practice

9. Moesha has 196 pepper plants that she wants to plant in a square formation. How many pepper plants should she plant in each row?

10. Equation Editor What is the value of p in the equation shown?

$$p^3 = -0.027$$

←	→	↶	↷	⌫

1	2	3	+	−	×	÷
4	5	6	<	≤	=	≥	>
7	8	9	$\frac{x}{n}$	x^n	( )				$\sqrt{x}$	$\sqrt[n]{x}$	π
0	.	−				

Apply

11. A cement path consists of six equal-sized cement squares arranged in a row to form a rectangle. If the total area of the path is 96 square feet, what is the length in feet of the path?

12. A photo collage consists of seven equal-sized square photos arranged in a row to form a rectangle. If the total area of the collage is 567 square inches, what is the length of the collage?

13. **(MP) Reason Inductively** Explain why $\sqrt[3]{8}$ is a rational number, but $\sqrt{8}$ is not a rational number.

14. Give an example of when the decimal equivalent of a square root would be rounded to an approximate value. Explain why it is appropriate to round.

15. Write a number that completes the analogy:

x^2 is to 441 as x^3 is to _____.

16. **(MP) Identify Repeated Reasoning** Simplify each expression. Then write a rule for the pattern.
   a. $(\sqrt{81})^2 =$ _____

   b. $\left(\sqrt{\dfrac{9}{16}}\right)^2 =$ _____

   c. $(\sqrt{0.04})^2 =$ _____

   d. $(\sqrt{t})^2 =$ _____

Real Numbers

I Can... identify irrational numbers and name the set(s) of real numbers to which a given real number belongs.

Explore Real Numbers

Online Activity You will use number lines to explore the set of real numbers.

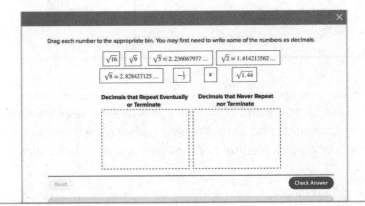

Drag each number to the appropriate bin. You may first need to write some of the numbers as decimals.

$\sqrt{16}$ $\sqrt{9}$ $\sqrt{5} \approx 2.236067977\ldots$ $\sqrt{2} \approx 1.414213562\ldots$

$\sqrt{8} \approx 2.828427125\ldots$ $-\frac{1}{3}$ π $\sqrt{1.44}$

Decimals that Repeat Eventually or Terminate

Decimals that Never Repeat nor Terminate

Reset   Check Answer

Learn Real Numbers

Real numbers are numbers that can be found on the number line.

Go Online Watch the animation to plot the numbers -4.5, -1, $\frac{3}{4}$, π, and $\sqrt{21}$ on the number line.

$$-5 \quad -4 \quad -3 \quad -2 \quad -1 \quad 0 \quad 1 \quad 2 \quad 3 \quad 4 \quad 5$$

Real numbers are either rational, with a decimal expansion that terminates or repeats, or irrational. An **irrational number** is a number that cannot be expressed as the ratio $\frac{a}{b}$, where a and b are integers and $b \neq 0$. Irrational numbers have decimal expansions that are non-terminating and non-repeating.

The square root of any number that is not a perfect square is irrational.

$\sqrt{3} \approx 1.732050808\ldots$ $\sqrt{5} \approx 2.2360679775\ldots$

(continued on next page)

💬 **Talk About It!**

What sets of numbers do the rational numbers include? Why is $\sqrt{2}$ an irrational number? π?

Write the following numbers in the Venn diagram.

$$-8, -3, -\frac{3}{4}, 0, 18\%, 0.\overline{4}, 0.45, \frac{1}{2}, \pi, 1.21231234..., \sqrt{3}, 4, 9, 10$$

Real Numbers

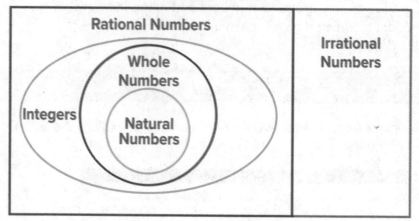

A rational number is a number that can be expressed as the ratio $\frac{a}{b}$, where a and b are integers and $b \neq 0$. Integers are the set of natural numbers, their opposites, and zero. Whole numbers are the set of natural numbers and zero. Natural numbers are the set of counting numbers.

When written as decimals, irrational numbers neither terminate, nor repeat eventually.

Example 1 Identify Real Numbers

Determine whether −25 is *rational* or *irrational*.

Can −25 be expressed as a ratio in the form $\frac{a}{b}$?

    Yes   No

When written in the form $\frac{a}{b}$, are a and b integers and $b \neq 0$?

    Yes   No

A rational number is a number that can be expressed as the ratio of two integers.

Since $-25 = \dfrac{\boxed{}}{\boxed{}}$, −25 is a rational number.

Check

Circle whether $-\frac{2}{9}$ is *rational* or *irrational*.

rational   irrational

Example 2 Classify Real Numbers

Name all the sets of numbers to which the real number 0.2525... belongs.

The set of real numbers include natural numbers, whole numbers, integers, rational numbers, and irrational numbers.

Can 0.2525... be expressed as a ratio in the form $\frac{a}{b}$, where a and b are integers and $b \neq 0$?

Yes   No

Is 0.2525... from the set of integers {..., –3, –2, –1, 0, 1, 2, 3, ...}?

Yes   No

So, 0.2525... is a rational number because it is equivalent to $\dfrac{\boxed{}}{\boxed{}}$.

Talk About It!

What is another way to write 0.2525...?

Check

To which set(s) of numbers does the real number $\frac{21}{\sqrt{4}}$ belong? Select all that apply.

☐ Rational

☐ Irrational

☐ Integer

☐ Whole

☐ Natural

Show your work here

Go Online You can complete an Extra Example online.

Example 3 Classify Real Numbers

Name all the sets of numbers to which the real number $\sqrt{36}$ belongs.

Can $\sqrt{36}$ be expressed as a ratio in the form $\frac{a}{b}$, where a and b are integers and $b \neq 0$?

    Yes   No

Is $\sqrt{36}$ from the set of integers {..., −3, −2, −1, 0, 1, 2, 3, ...}?

    Yes   No

Is $\sqrt{36}$ from the set of whole numbers {0, 1, 2, 3, ...}?

    Yes   No

Is $\sqrt{36}$ from the set of natural numbers {1, 2, 3, ...}?

    Yes   No

Since $\sqrt{36} =$ _____, it is a natural number, a whole number, an integer, and a rational number.

Check

To which set(s) of numbers does the real number $-\sqrt{64}$ belong? Select all that apply.

☐ Rational

☐ Irrational

☐ Integer

☐ Whole

☐ Natural

(Show your work here)

🅑 **Go Online** You can complete an Extra Example online.

Example 4 Classify Real Numbers

Name all the sets of numbers to which the real number $-\sqrt{7}$ belongs.

$$-\sqrt{7} \approx -2.645751311...$$

Does the decimal terminate? repeat eventually?

Yes   No

Can $-\sqrt{7} \approx -2.645751311...$ be expressed as a ratio in the form $\frac{a}{b}$, where a and b are integers and $b \neq 0$?

Yes   No

The decimal value of $-\sqrt{7}$ neither terminates nor repeats eventually, so it is an irrational number.

Check

To which set(s) of numbers does the real number π belong? Select all that apply.

☐ Rational

☐ Irrational

☐ Integer

☐ Whole

☐ Natural

Show your work here

Go Online You can complete an Extra Example online.

Think About It!
Can you simplify the square root?

Talk About It!
How can you tell, just by studying the expression, that $-\sqrt{7}$ is irrational?

Learn Describe Sets of Real Numbers

Some sets of numbers are subsets of other sets of numbers. For example, rational numbers and irrational numbers are subsets of real numbers.

A Venn diagram can be used to describe the relationship between sets of real numbers.

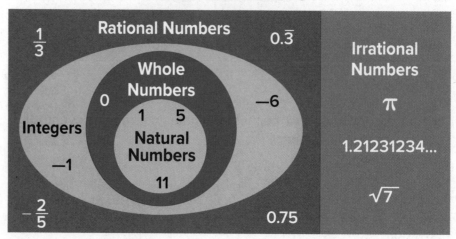

Real Numbers

Go Online Watch the animation, or use the Venn diagram, to complete the following sentences.

_____ numbers are a subset of whole numbers.

_____ numbers are subsets of integers.

Integers are a subset of _____ numbers.

Rational numbers and irrational numbers are subsets of _____ numbers.

If a given statement about real numbers is false, you can provide a **counterexample**, which is a statement or example that shows a conjecture is false.

Pause and Reflect

Describe the decimal form of irrational numbers. How are they represented in the Venn diagram?

Talk About It!

Natural numbers are a subset of whole numbers. What other subsets of numbers are shown in the Venn diagram?

Example 5 Describe Sets of Real Numbers

Use the Venn diagram to determine whether the statement is *true* or *false*. If the statement is *true*, explain your reasoning. If the statement is *false*, provide a counterexample.

All rational numbers are integers.

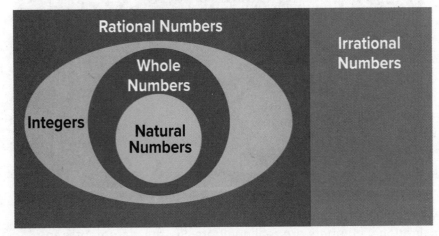

Part A Determine whether the statement is *true* or *false*.

Integers are a subset of _____ numbers. So, the statement is false.

Part B Provide a counterexample.

One possible counterexample is 0.6. The decimal 0.6 is a rational number, but not an integer.

Check

Use the Venn diagram above to determine whether the statement is *true* or *false*. If the statement is *true*, explain your reasoning. If the statement is *false*, provide a counterexample.

All whole numbers are natural numbers.

Part A

Determine whether the statement is *true* or *false*.

Part B

If the statement is *true*, explain your reasoning. If the statement is *false*, provide a counterexample.

Show your work here

Go Online You can complete an Extra Example online.

Example 6 Describe Sets of Real Numbers

Use the Venn diagram to determine whether the statement is *true* or *false*. If the statement is *true*, explain your reasoning. If the statement is *false*, provide a counterexample.

All irrational numbers are real numbers.

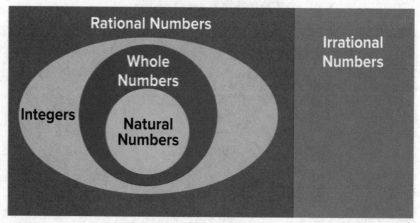

By definition, irrational numbers are a subset of _____ numbers. So, irrational numbers are real numbers. So, the statement is true.

Check

Use the Venn diagram above to determine whether the statement is *true* or *false*.

All natural numbers are whole numbers.

Part A
Determine whether the statement is *true* or *false*.

Part B
If the statement is *true*, explain your reasoning. If the statement is *false*, provide a counterexample.

📖 **Foldables** It's time to update your Foldable, located in the Module Review, based on what you learned in this lesson. If you haven't already assembled your Foldable, you can find the instructions on page FL1.

Practice

Go Online You can complete your homework online.

Identify whether each number is *rational* or *irrational*. (Example 1)

1. $-\sqrt{10}$

2. $-\dfrac{3}{11}$

3. $0.\overline{3}$

4. $\sqrt{81}$

5. 0

6. $-\dfrac{\sqrt{2}}{2}$

7. $\sqrt{7}$

8. $\dfrac{\sqrt{2}}{\sqrt{2}}$

Select all the sets of numbers to which each real number belongs. (Examples 2–4)

9. $\sqrt[3]{343}$

Ⓐ Rational

Ⓑ Irrational

Ⓒ Integer

Ⓓ Whole

Ⓔ Natural

10. $\dfrac{7}{\sqrt{2}}$

Ⓐ Rational

Ⓑ Irrational

Ⓒ Integer

Ⓓ Whole

Ⓔ Natural

11. $-\dfrac{7}{1}$

Ⓐ Rational

Ⓑ Irrational

Ⓒ Integer

Ⓓ Whole

Ⓔ Natural

Determine whether each statement is *true* or *false*. If the statement is *true*, explain your reasoning. If the statement is *false*, provide a counterexample. (Examples 5 and 6)

12. A number cannot be irrational and an integer.

13. All integers are rational.

14. **Multiselect** Select the numbers that are part of the set of rational numbers.

☐ $-\dfrac{11}{\sqrt{9}}$

☐ $\dfrac{1}{\sqrt{2}}$

☐ $\sqrt{-16}$

☐ $\dfrac{\sqrt[3]{-4096}}{\sqrt{16}}$

☐ $\sqrt[3]{4}$

☐ $0.333\ldots$

15. **MP** **Use Math Tools** Explain how you could use a calculator to determine if $\sqrt{8}$ expressed as a decimal ever terminates.

MP **Justify Conclusions** Determine whether each statement is *true* or *false*. If the statement is false, give a counterexample or explain your reasoning.

16. The product of a non-zero rational number and an irrational number is rational.

17. Expressing $\sqrt{2}$ as the ratio $\dfrac{\sqrt{2}}{1}$ means that $\sqrt{2}$ is a rational number.

18. The product of two irrational numbers is irrational.

Estimate Irrational Numbers

I Can... estimate irrational numbers by approximating their locations on a number line or by truncating their decimal expansions.

Explore Roots of Non-Perfect Squares

Online Activity You will use Web Sketchpad to explore how to find the square root of a non-perfect square.

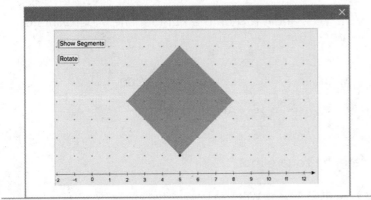

Learn Estimate Irrational Numbers Using a Number Line

The decimal expansion of an irrational number never repeats nor terminates. To write the decimal expansion of an irrational number, you can denote an approximation using the $\approx$ symbol. Use more place values to give more accurate approximations.

Go Online Watch the animation to see how to estimate $\sqrt{8}$ using a number line using the following steps:

Step 1 Since 8 is between two perfect squares, 4 and 9,
$\sqrt{4} < \sqrt{8} < \sqrt{9}$.

Step 2 Graph $\sqrt{4} = 2$ and $\sqrt{9} = 3$ on the number line.

Step 3 Since 8 is closer to 9 than to 4, graph $\sqrt{8}$ closer to $\sqrt{9}$ than to $\sqrt{4}$.

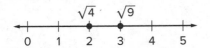

Talk About It!
How do you know that $\sqrt{8}$ is between 2 and 3? How do you know that $\sqrt{8}$ is closer to 3 than 2?

 Think About It!

How would you begin estimating the square root? What are some perfect squares that are close to 83?

Example 1 Estimate Square Roots to the Nearest Integer

Estimate $\sqrt{83}$ to the nearest integer.

Step 1 Find two perfect squares between which 83 lies. Find their square roots.

The greatest perfect square less than 83 is _____, and $\sqrt{81}$ = _____.

The least perfect square greater than 83 is _____, and $\sqrt{100}$ = _____.

Step 2 Plot $\sqrt{81}$, $\sqrt{83}$, and $\sqrt{100}$ on the number line. Approximate the location of $\sqrt{83}$.

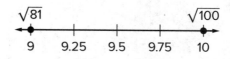

Step 3 Estimate the square root.

$81 < 83 < 100$ Write an inequality.

$9^2 < 83 < 10^2$ $81 = 9^2$ and $100 = 10^2$

$\sqrt{9^2} < \sqrt{83} < \sqrt{10^2}$ Find the square root of each number.

$\boxed{} < \sqrt{83} < \boxed{}$ Simplify.

So, $\sqrt{83}$ is between 9 and 10. Since $\sqrt{83}$ is closer to _____ on a number line, the best integer estimate for $\sqrt{83}$ is 9.

Check

Estimate $\sqrt{135}$ to the nearest integer.

 Go Online You can complete an Extra Example online.

Example 2 Estimate Square Roots to the Nearest Tenth

Estimate $\sqrt{83}$ to the nearest tenth.

Step 1 Determine the integer closest to $\sqrt{83}$.

$\sqrt{83}$ is much closer to $\sqrt{81}$, or _____, than it is to $\sqrt{100}$, or _____.

Step 2 Test intervals close to 9. Start with the interval between 9 and 9.1.

$9 < \sqrt{83} < 9.1$ Write an inequality.

$9^2 < (\sqrt{83})^2 < 9.1^2$ Square the values.

$\boxed{} \overset{?}{<} 83 \overset{?}{<} \boxed{}$ Simplify. Is the inequality true?

The inequality is not true because 83 is not between 81 and 82.81.

Step 3 Test the next interval, 9.1 to 9.2.

$9.1 < \sqrt{83} < 9.2$ Write an inequality.

$9.1^2 < (\sqrt{83})^2 < 9.2^2$ Square the values.

$\boxed{} \overset{?}{<} 83 \overset{?}{<} \boxed{}$ Simplify. Is the inequality true?

The inequality is true because 83 is between 82.81 and 84.64.

So, $\sqrt{83}$ is between 9.1 and 9.2. Since 83 is closer to _____ than it is to 84.64, $\sqrt{83} \approx 9.1$.

Check

Estimate $\sqrt{106}$ to the nearest tenth.

 Show your work here

 Go Online You can complete an Extra Example online.

Think About It!

Between what two integers does $\sqrt{83}$ lie on a number line?

Talk About It!

How can you use a graph to verify that $\sqrt{83}$ is closer to 9.1 than 9.2? How could you continue on to get a better approximation of $\sqrt{83}$?

Example 3 Estimate Cube Roots to the Nearest Integer

Estimate $\sqrt[3]{320}$ to the nearest integer.

Step 1 Find two perfect cubes between which 320 lies. Find their cube roots.

The greatest perfect cube less than 320 is _____, and $\sqrt[3]{216} = $ _____.

The least perfect cube greater than 320 is _____, and $\sqrt[3]{343} = $ _____.

Step 2 Plot $\sqrt[3]{216}$, $\sqrt[3]{320}$, and $\sqrt[3]{343}$ on the number line. Approximate the location of $\sqrt[3]{320}$.

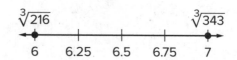

Step 3 Estimate the cube root.

$$216 < \ 320 \ < 343 \qquad \text{Write an inequality.}$$
$$6^3 < \ 320 \ < 7^3 \qquad 216 = 6^3 \text{ and } 343 = 7^3$$
$$\sqrt[3]{6^3} < \sqrt[3]{320} < \sqrt[3]{7^3} \qquad \text{Find the cube root of each number.}$$
$$\boxed{} < \sqrt[3]{320} < \boxed{} \qquad \text{Simplify.}$$

So, $\sqrt[3]{320}$ is between 6 and 7. Since 320 is closer to _____, the best integer estimate for $\sqrt[3]{320}$ is 7.

Check

Estimate $\sqrt[3]{51}$ to the nearest integer.

Show your work here

Go Online You can complete an Extra Example online.

Learn Estimate Irrational Numbers by Truncating

You can estimate irrational numbers using a calculator by truncating the decimal expansion. **Truncating** is a process of approximating a decimal number by eliminating all decimal places past a certain point without rounding.

Truncate $\sqrt{12} \approx 3.464101615...$ to the specified decimal places.

Tenths

$\sqrt{12} \approx 3.4$~~64101615~~...

Truncate, or drop, the digits after the tenths place.

Hundredths

$\sqrt{12} \approx$ _____

Truncate, or drop, the digits after the hundredths place.

Thousandths

$\sqrt{12} \approx$ _____

Truncate, or drop, the digits after the thousandths place.

💬 **Talk About It!**

How could you continue truncating $\sqrt{12}$ to get better approximations?

💬 **Talk About It!**

If you *rounded* $\sqrt{12}$ to the nearest tenth, what would be the approximation?

If you *truncated* $\sqrt{12}$ to the nearest tenth, what would be the approximation?

What is the difference between truncating and rounding?

Pause and Reflect

Explain how truncating can be used to estimate the value of π.

Record your observations here

Think About It!

Between which two whole numbers is $\sqrt{2}$?

Talk About It!

Why will Wyatt need more than 5.6 meters, but less than 6.0 meters of fencing?

Talk About It!

Could you truncate the decimal expansion differently? Explain how it would affect the answer.

🌐 Example 4 Estimate by Truncating

Wyatt wants to fence in a square portion of the yard to make a play area for his new puppy. The area covered is 2 square meters.

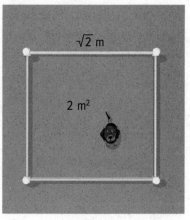

How much fencing should he buy?

The amount of fencing he will need is the _____ of the square, $4 \cdot \sqrt{2}$ meters.

Step 1 Approximate $4\sqrt{2}$ by first truncating the decimal expansion of $\sqrt{2}$ to the tenths place.

$\sqrt{2} \approx$ _____   Use a calculator.

$\sqrt{2} \approx$ _____   Truncate the digits after the tenths place.

So, $\sqrt{2}$ is between 1.4 and 1.5.

Step 2 Find the amount of fencing.

Since $\sqrt{2}$ is between 1.4 and 1.5, the perimeter, $4(\sqrt{2})$, is between 4(1.4) and 4(1.5).

$4(1.4) < 4(\sqrt{2}) < 4(1.5)$ Write the inequality.

$\boxed{} < 4(\sqrt{2}) < \boxed{}$ Multiply.

So, Wyatt will need between 5.6 and 6.0 meters of fencing. Therefore, he should buy 6 meters of fencing.

Check

Tobias dropped a tennis ball from a height of 60 feet. The time in seconds it takes for the ball to fall 60 feet is found using the expression $0.25 \cdot \sqrt{60}$. Determine the number of seconds it takes for the ball to fall 60 feet. Truncate the value of $\sqrt{60}$ to the tenths place.

Show your work here

🐢 **Go Online** You can complete an Extra Example online.

 Apply Golden Rectangle

The *golden rectangle* can be seen in the structure of a nautilus shell. The ratio of the longer side length to the shorter side is equal to $\dfrac{1 + \sqrt{5}}{2}$. Estimate this value.

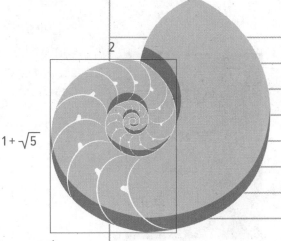

1 What is the task?

Make sure you understand exactly what question to answer or problem to solve. You may want to read the problem three times. Discuss these questions with a partner.

First Time Describe the context of the problem, in your own words.
Second Time What mathematics do you see in the problem?
Third Time What are you wondering about?

2 How can you approach the task? What strategies can you use?

3 What is your solution?

Use your strategy to solve the problem.

4 How can you show your solution is reasonable?

✏️ **Write About It!** Write an argument that can be used to defend your solution.

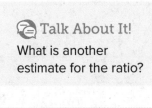
Talk About It!
What is another estimate for the ratio?

Check

In Little League, the bases are squares with sides of 14 inches. The expression $\sqrt{s^2 + s^2}$ represents the distance *diagonally* across a square of side length s. Estimate the diagonal distance across a base to the nearest inch.

Show your work here

Go Online You can complete an Extra Example online.

Pause and Reflect

Compare and contrast estimating rational numbers on a number line with estimating irrational numbers on a number line.

Record your observations here

Name ______________________ Period ____________ Date ____________

Practice

Go Online You can complete your homework online.

Estimate each square root or cube root to the nearest integer. **(Examples 1 and 3)**

1. $\sqrt{125} \approx$ ________

2. $\sqrt{55} \approx$ ________

3. $\sqrt[3]{70} \approx$ ________

4. $\sqrt[3]{923} \approx$ ________

Estimate each square root to the nearest tenth. **(Example 2)**

5. $\sqrt{296} \approx$ ________

6. $\sqrt{5} \approx$ ________

7. $\sqrt{11} \approx$ ________

8. $\sqrt{62} \approx$ ________

9. The formula $s = \sqrt{18d}$ can be used to find the speed s of a car in miles per hour when the car needs d feet to come to a complete stop after stepping on the brakes. If it took a car 25 feet to come to a complete stop after stepping on the brakes, estimate the speed of the car. Truncate the value of $\sqrt{18d}$, when $d = 25$, to the tenths place. **(Example 4)**

Test Practice

10. If the area of a square is 32 square feet, estimate the length of each side of the square to the nearest whole number.

11. Equation Editor Estimate the square root to the nearest tenth.

$\sqrt{489}$

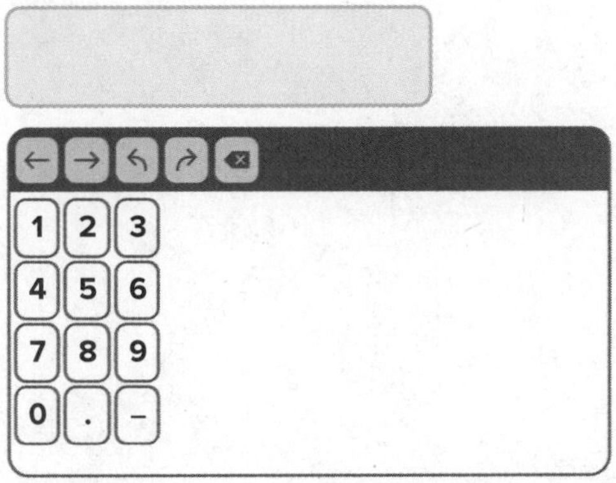

Apply

12. The formula $t = \dfrac{\sqrt{h}}{4}$ represents the time t in seconds that it takes an object to fall from a height of h feet. If a rock falls from 125 feet, estimate how long it will take the rock to hit the ground. Estimate the square root to the nearest integer.

13. The radius of a circle with area A can be approximated using the formula $r = \sqrt{\dfrac{A}{3}}$. Estimate the radius of a wrestling mat circle with an area of 452 square feet. Round to the nearest Integer.

Higher-Order Thinking Problems

14. **MP Find the Error** A classmate estimated $\sqrt{397}$ to be about 200. Explain the mistake and correct it.

15. **MP Be Precise** Explain how to write the exact value for the square root of a non-perfect square. Give an example.

16. Carrie is packing her clothes in moving boxes that are in the shape of a cube. Each box has a volume of 3 cubic feet. The moving truck has shelves that are 12 inches in height. Will the moving boxes fit? Explain.

17. **MP Make an Argument** Explain how you could estimate $\sqrt[4]{20}$ to the nearest integer.

Compare and Order Real Numbers

I Can... use rational approximations to compare and order real numbers, including irrational numbers.

Learn Compare and Order Real Numbers

You can compare and order real numbers by writing them in the same form. One way to do this is to use or approximate the decimal expansion of each number in order to compare or order a set of numbers.

Complete the following to compare and order the set of numbers shown.

$$\frac{18}{5}, \pi, \sqrt{11}$$

Write each number in decimal notation.

$\frac{18}{5} = $ ☐ $\pi \approx$ ☐ $\sqrt{11} \approx$ ☐

Compare each set of numbers using <, >, or =.

3.14 ☐ 3.32 ☐ 3.6

π ☐ $\sqrt{11}$ ☐ $\frac{18}{5}$

Graph each number on the number line.

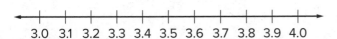

3.0 3.1 3.2 3.3 3.4 3.5 3.6 3.7 3.8 3.9 4.0

Order the set of numbers from least to greatest.

——— , ——— , ———

Think About It!

Between which two integers does $\sqrt{8}$ lie? To which integer is it closer?

Talk About It!

Is $\sqrt{8}$ closer to 2.8 or 2.9? Explain.

Example 1 Compare Real Numbers

Which symbol, <, >, or =, would complete the statement
$\sqrt{8}$ _____ $2\frac{2}{3}$ **to make a true statement? Then graph the numbers on a number line.**

Part A Compare the numbers.

Approximate the decimal expansion of each number.

$\sqrt{8} \approx$ ☐   Estimate to the nearest tenth.

$2\frac{2}{3} =$ ☐   Write using bar notation.

Since 2.8 is greater than $2.\overline{6}$, $\sqrt{8} > 2\frac{2}{3}$.

Part B Graph the numbers on the number line.

Approximate the location of each number.

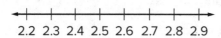

2.2 2.3 2.4 2.5 2.6 2.7 2.8 2.9

Pause and Reflect

When you first saw this Example, what was your reaction? Did you think you could solve the problem? Did what you already know help you solve the problem?

> Record your observations here

Check

Which symbol, <, >, or =, would complete the statement
$\frac{\sqrt{25}}{2}$ _____ $\sqrt{6.25}$ to make a true statement? Then graph the
numbers on a number line.

Part A

Write the symbol, <, >, or =, that makes $\frac{\sqrt{25}}{2}$ _____ $\sqrt{6.25}$ a true
statement.

Part B

Which of the following is the correct graph of $\frac{\sqrt{25}}{2}$ and $\sqrt{6.25}$ on a
number line?

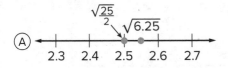

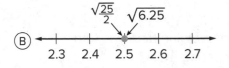

Ⓒ

Ⓓ

🅑 **Go Online** You can complete an Extra Example online.

Think About It!

How would you begin comparing the numbers?

Example 2 Compare Real Numbers

Which symbol, <, >, or =, would complete the statement $-\sqrt{6}$ _____ $-\frac{\pi}{2}$ to make a true statement? Then graph the numbers on a number line.

Part A Compare the numbers.

Approximate the decimal expansion of each number.

$-\sqrt{6} \approx$ [ ]　　　　　　　　Estimate to the nearest tenth.

$-\frac{\pi}{2} \approx -\frac{3.14}{2}$ or [ ]　　　Estimate to the nearest hundredth.

Since -2.4 is less than -1.57, $-\sqrt{6} < -\frac{\pi}{2}$.

Part B Graph the numbers on the number line.

Approximate the location of each number. When graphing numbers on the number line, greater numbers are graphed farther to the right.

Pause and Reflect

How does graphing numbers on a number line help you know whether to use <, >, or = when comparing them?

> Record your observations here

Check

Which symbol, $<$, $>$, or $=$, would complete the statement
$-\sqrt{5}$ _____ -210% to make a true statement? Then graph the
numbers on a number line.

Part A

Write the symbol, $<$, $>$, or $=$, that makes $-\sqrt{5}$ _____ -210% a
true statement.

Part B

Which of the following is the correct graph of $-\sqrt{5}$ and -210% on a
number line?

(A)
$-\sqrt{5}$ -210%

-2.4 -2.3 -2.2 -2.1 -2.0

(B)
$-\sqrt{5}$ -210%

-2.4 -2.3 -2.2 -2.1 -2.0

(C)
$-\sqrt{5}$ -210%

-2.4 -2.3 -2.2 -2.1 -2.0

(D)
$-\sqrt{5}$ -210%

-2.4 -2.3 -2.2 -2.1 -2.0

 **Go Online** You can complete an Extra Example online.

Example 3 Order Real Numbers

Order the set $\left\{\sqrt{30}, 6, 5\frac{4}{5}, 5.3\overline{6}\right\}$ **from least to greatest. Then graph the set on the number line.**

Part A Order the set of numbers.

Approximate the decimal expansion of each number.

$\sqrt{30} \approx$ [   ]   Estimate to the nearest tenth.

$6 = 6.00$ Write as a decimal.

$5\frac{4}{5} =$ [   ]   Write as a decimal.

$5.3\overline{6} \approx$ [   ]   Write as a decimal to the nearest hundredth.

Write the decimals from least to greatest.

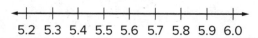
_____ , _____ , _____ , _____

So, from least to greatest, the order is $5.3\overline{6}$, $\sqrt{30}$, $5\frac{4}{5}$, and 6.

Part B Graph the numbers on the number line.

Approximate the location of $\sqrt{30}$ and $5.3\overline{6}$.

5.2 5.3 5.4 5.5 5.6 5.7 5.8 5.9 6.0

Pause and Reflect

How are comparing real numbers and ordering real numbers related to each other?

Record your observations here

🍩 **Think About It!**

How would you begin ordering the numbers?

Check

Order the set $\left\{\sqrt{3}, \frac{\pi}{2}, 160\%, 1.3\overline{5}\right\}$ from least to greatest. Then graph the set on the number line.

Part A

Which of the following is the correct order of the set from least to greatest?

Ⓐ $\left\{1.3\overline{5}, \frac{\pi}{2}, 160\%, \sqrt{3}\right\}$

Ⓑ $\left\{1.3\overline{5}, \sqrt{3}, 160\%, \frac{\pi}{2}\right\}$

Ⓒ $\left\{1.3\overline{5}, 160\%, \frac{\pi}{2}, \sqrt{3}\right\}$

Ⓓ $\left\{\sqrt{3}, 1.3\overline{5}, 160\%, \frac{\pi}{2}\right\}$

Show your work here

Part B

Which of the following is the correct graph of the set of numbers?

Ⓐ
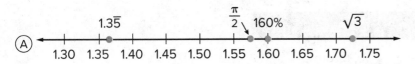

Ⓑ
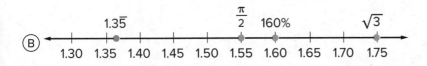

Ⓒ

Ⓓ
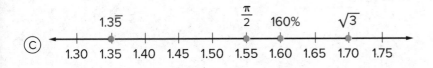

 **Go Online** You can complete an Extra Example online.

On Wednesday, there is an $83\frac{1}{3}$% chance of rain. On Thursday, there is a $\frac{9}{10}$ chance of rain. On Friday, there is a **6 out of 7** chance that it will rain.

On which day is there the greatest chance of rain?

Step 1 Write each number in decimal notation. Round to the nearest hundredth.

$$83\frac{1}{3}\% \approx \boxed{} \qquad \frac{9}{10} = \boxed{} \qquad 6 \text{ out of } 7 \approx \boxed{}$$

Step 2 Order the decimals.

Since $0.9 > 0.86 > 0.83$, then $\frac{9}{10} \boxed{} \frac{6}{7} \boxed{} 83\frac{1}{3}\%$.

So, there is the greatest chance it will rain on Thursday.

Check

The table shows the on-base statistics for three players at a recent baseball tournament. Which player had the greatest on-base statistic?

Player	On-Base Statistic
1	15 out of 21
2	$\frac{14}{19}$
3	72.5%

Show your work here

🌐 **Go Online** You can complete an Extra Example online.

🌐 **Apply** Line of Sight

On a clear day, the number of miles a person can see to the horizon is about $1.23\sqrt{h}$, where h is the person's height from the ground in feet. Suppose Frida is at the Empire State Building observation deck at 1,050 feet and Logan is at the Freedom Tower observation deck at 1,254 feet. How much farther can Logan see than Frida from the observation deck?

🔲 Go Online
Watch the animation.

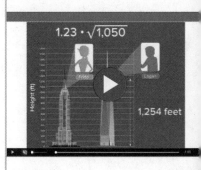

$$1.23 \cdot \sqrt{1,050}$$

1,254 feet

1 What is the task?

Make sure you understand exactly what question to answer or problem to solve. You may want to read the problem three times. Discuss these questions with a partner.

First Time Describe the context of the problem, in your own words.
Second Time What mathematics do you see in the problem?
Third Time What are you wondering about?

2 How can you approach the task? What strategies can you use?

Record your observations here

3 What is your solution?

Use your strategy to solve the problem.

Show your work here

🗨 Talk About It!

How could you use the nearest perfect squares to check for reasonableness?

4 How can you show your solution is reasonable?

🔵 **Write About It!** Write an argument that can be used to defend your solution.

Check

The time in seconds that it takes an object to fall d feet can be found using the expression $\frac{\sqrt{d}}{4}$. Suppose Aiden drops a tennis ball from a height of 50 feet at the same time Mason drops a similar tennis ball from a height of 20 feet. How much longer will it take Aiden's tennis ball to reach the ground than Mason's tennis ball? Round to the nearest hundredth.

Go Online You can complete an Extra Example online.

Pause and Reflect

Review the Examples from this module. Which one did you find most challenging? What are the steps you would take to solve a problem of this type?

Practice

Go Online You can complete your homework online.

Complete each statement using <, >, or =. Then graph the numbers on the number line. (Examples 1 and 2)

1. $\sqrt{11}$ _____ $3\frac{2}{3}$

2. $\sqrt{3}$ _____ $\frac{\sqrt{10}}{2}$

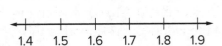

3. $-\pi^2$ _____ $-\sqrt{93}$

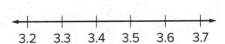

4. $-\sqrt{12}$ _____ -320%

5. Order the set $\left\{3\frac{1}{2}, \frac{10}{3}, \pi, \sqrt{13}\right\}$ from least to greatest. Then graph the set on the number line. (Example 3)

---

6. The table shows the foul-shot statistics for three players in a recent basketball game. Which player had the greatest foul-shot statistic? (Example 4)

Player	Foul-Shot Statistic
1	$\frac{7}{9}$
2	72%
3	8 out of 10

Test Practice

7. Multiple Choice Select the symbol that makes the sentence true.

$$\sqrt{27} \underline{\quad} \frac{\sqrt{95}}{2}$$

Ⓐ <

Ⓑ >

Ⓒ =

Ⓓ ≤

Apply

8. The radius of a circle can be approximated using the expression $\sqrt{\frac{A}{3}}$ where A represents the area. A circular kiddie swimming pool has an area of about 28 square feet. An inflatable full-size circular pool has an area of about 113 square feet. How much greater is the radius of the full-size pool than the radius of the kiddie pool? Round to the nearest whole number.

9. The time in seconds that it takes an object to fall d feet can found using the expression $\frac{\sqrt{d}}{4}$. In an egg drop contest, Clara successfully dropped her egg container from a height of 35 feet, while Vladimir successfully dropped his egg container from a height of 23 feet. How much longer did it take Clara's egg to reach the floor than Vladimir's egg? Round to the nearest tenth.

10. **Which One Doesn't Belong?** Identify the number that does not belong in the group. Explain your reasoning.

 $-23.\overline{2}$ $-23\frac{1}{5}$ $\boxed{-\sqrt{23}}$ $\boxed{-23.2}$

11. **MP** **Justify Conclusions** Which number is greater, 3.14 or π? Justify your answer.

12. **MP** **Find the Error** Kendra states that $\sqrt{3} > 2$ because 3 is greater than 2. Explain Kendra's mistake and correct it.

13. Identify two numbers, one rational and one irrational, that are between 1.6 and 1.8. Write an inequality to compare the two numbers.

📙 **Foldables** Use your Foldable to help review the module.

Real Numbers

Examples	Examples

Rate Yourself! ⬛ ◆ ★

Complete the chart at the beginning of the module by placing a checkmark in each row that corresponds with how much you know about each topic after completing this module.

Write about one thing you learned.

_____

_____

_____

_____

_____

_____

Write about a question you still have.

_____

_____

_____

_____

_____

_____

Reflect on the Module

Use what you learned about real numbers to complete the graphic organizer.

ⓔ Essential Question

Why do we classify numbers?

Real Number

What is it?

Rational Number

What is it?

Examples

Real-World Situations

Irrational Number

What is it?

Examples

Real-World Situations

Name _____   Period _____   Date _____

Test Practice

1. Open Response The Jefferson football teams played several games this season. Each team's record is shown in the table. **(Lesson 1)**

Football Team	Wins	Losses
Varsity	8	3
Junior Varsity	7	3

A. Find the winning average of each team written as a decimal. Then determine if the winning averages are terminating decimals.

B. Which team had a better season?

2. Multiple Choice Which decimal is equivalent to $-3\frac{19}{30}$? **(Lesson 1)**

Ⓐ -3.63

Ⓑ $-3.6\overline{3}$

Ⓒ $-3.\overline{63}$

Ⓓ $-3.6363...$

3. Equation Editor A local store sells bags of game pieces with various colors. Four out of every 19 pieces are red. The store sold a bag with 76 game pieces. What number of the game pieces are red? **(Lesson 1)**

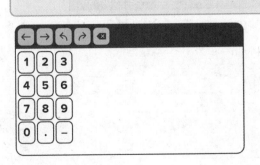

4. Multiselect Simplify $\pm\sqrt{1.69}$. Select all that apply. **(Lesson 2)**

☐ -0.13

☐ 0.13

☐ -1.3

☐ 1.3

☐ 13

☐ -13

5. Open Response A vertical shelving unit consists of five equal-sized square shelves arranged in a column to form a rectangle. The total area of all five shelves is 500 square inches. **(Lesson 2)**

A. What is the height, in inches, of the shelving unit?

B. Explain how to find the height of the shelving unit.

6. Multiselect Simplify $-\sqrt{\frac{81}{121}}$. Select all that apply. **(Lesson 2)**

☐ $-0.66942...$

☐ $\frac{9}{11}$

☐ $-\frac{9}{11}$

☐ $-0.8181...$

☐ $0.8181...$

7. Table Item Indicate whether each real number is rational or irrational. (Lesson 3)

	Rational	Irrational
-6		
$\sqrt{7}$		
$3\frac{1}{2}$		

8. Open Response Use the Venn diagram. (Lesson 3)

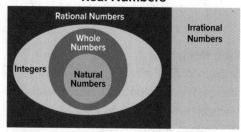

Real Numbers

A. Determine whether the statement is *true* or *false*.

All integers are natural numbers.

B. If the statement is *true*, explain your reasoning. If the statement is *false*, provide a counterexample.

9. Table Item Indicate the integer to which each square root is closest on a number line. (Lesson 4)

	7	8	9
$\sqrt{70}$			
$\sqrt{79}$			
$\sqrt{88}$			
$\sqrt{52}$			
$\sqrt{60}$			
$\sqrt{47}$			
$\sqrt{65}$			

10. Equation Editor A shipping box, in the shape of a cube, has a volume of 2,300 cubic inches. Estimate the length of the side of the shipping box to the nearest integer. (Lesson 4)

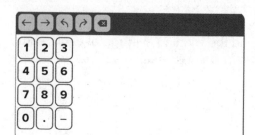

11. Open Response Consider the real numbers $-2\frac{1}{5}$ and $-\sqrt{7}$. (Lesson 5)

A. Compare the numbers. Use $<$, $>$, or $=$.
$-2\frac{1}{5}$ ___ $-\sqrt{7}$

B. Graph $-\sqrt{7}$ and $-2\frac{1}{5}$ on the number line.

12. Open Response The table shows the number of aces per serving attempts for three players at a recent volleyball tournament. Order the players from least aces per serving attempt to greatest aces per serving attempt. (Lesson 5)

Player	Number of Aces
Angela	9 out of 22
Jaylin	$22\frac{2}{9}\%$
Mya	$\frac{3}{10}$

Module 3

Solve Equations with Variables on Each Side

e Essential Question

How can equations with variables on each side be used to represent everyday situations?

What Will You Learn?

Place a checkmark (✓) in each row that corresponds with how much you already know about each topic **before** starting this module.

KEY ◼ — I don't know. ◆ — I've heard of it. ★ — I know it!	Before			After		
	◼	◆	★	◼	◆	★
solving equations with variables on each side						
writing equations with variables on each side						
solving multi-step equations						
writing multi-step equations						
solving equations with infinitely many solutions						
solving equations with no solution						
creating equations with infinitely many solutions						
creating equations with no solution						

Foldables Cut out the Foldable and tape it to the Module Review at the end of the module. You can use the Foldable throughout the module as you learn about solving equations with variables on each side.

What Vocabulary Will You Learn?

Check the box next to the vocabulary term if you may already know about it.

☐ like terms

Are You Ready?

Study the Quick Review to see if you are ready to start this module.
Then complete the Quick Check.

Quick Review

Example 1

Solve one-step addition and subtraction equations.

Solve $34 = x - 12$.

$34 = x - 12$	Write the equation.
$\underline{+12 = \quad +12}$	Addition Property of Equality
$46 = x$	Simplify.

Example 2

Solve one-step multiplication and division equations.

Solve $-5n = 35$.

$-5n = 35$	Write the equation.
$\dfrac{-5n}{-5} = \dfrac{35}{-5}$	Division Property of Equality
$n = -7$	Simplify.

Quick Check

1. Ana has 8 stamps. Together, Ricky and Ana have 23 stamps. The equation $23 = r + 8$ represents this situation, where r is the number of stamps Ricky has. Solve the equation to find the number of stamps Ricky has.

2. A certain number of plates will be placed on 8 tables. Each table will have 6 plates. The equation $\dfrac{p}{8} = 6$ represents this situation, where p is the total number of plates. Solve the equation to find the total number plates.

How Did You Do?

Which exercises did you answer correctly in the Quick Check?
Shade those exercise numbers at the right.

Solve Equations with Variables on Each Side

I Can... use the properties of equality to solve equations with variables on each side that have rational coefficients.

Explore Equations with Variables on Each Side

Online Activity You will use Web Sketchpad to explore how using a balance can help you solve equations with variables on each side.

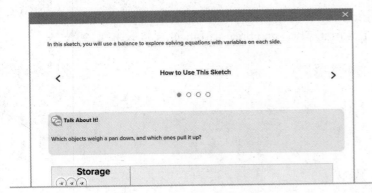

In this sketch, you will use a balance to explore solving equations with variables on each side.

< **How to Use This Sketch** >

● ○ ○ ○

Talk About It!

Which objects weigh a pan down, and which ones pull it up?

Storage

Learn Equations with Variables on Each Side

Some equations, like $8 + 3x = 5x + 2$, have variables on each side of the equals sign. To solve, use the properties of equality to write an equivalent equation with variables on one side of the equals sign. Then solve the equation.

$$8 + 3x = 5x + 2 \qquad \text{Write the equation.}$$
$$\underline{-3x = -3x} \qquad \text{Subtraction Property of Equality}$$
$$8 \quad = 2x + 2 \qquad \text{Simplify.}$$
$$\underline{-2 \quad = \quad -2} \qquad \text{Subtraction Property of Equality}$$
$$\frac{6}{2} \quad = \quad \frac{2x}{2} \qquad \text{Division Property of Equality}$$
$$3 = x \qquad \text{Simplify.}$$

(continued on next page)

Go Online Watch the animation to complete the steps to solve the equation $8 + 3x = 5x + 2$ using algebra tiles.

Step 1 Model the equation. Draw eight 1-tiles and three x-tiles to model the $8 + 3x$ on the left side of the mat. Draw five x-tiles and two 1-tiles to model $5x + 2$ on the right side of the mat.

Step 2 Draw the remaining tiles after removing three x-tiles from each side of the mat. The resulting equation is $8 = 2x + 2$.

Step 3 Draw the remaining tiles after removing two 1-tiles from each side of the mat. The resulting equation is $6 = 2x$.

Step 4 Draw the tiles so that the tiles are separated into two equal groups. There are three 1-tiles in each group, so $x = 3$.

Example 1 Solve Equations with Variables on Each Side

Solve $6n - 1 = 4n - 5$. Check your solution.

$$6n - 1 = 4n - 5 \qquad \text{Write the equation.}$$

$$\underline{-4n \qquad = \boxed{}} \qquad \text{Subtraction Property of Equality}$$

$$\boxed{} - 1 = -5 \qquad \text{Simplify.}$$

$$\underline{+1 = +1} \qquad \text{Addition Property of Equality}$$

$$2n = \boxed{} \qquad \text{Simplify.}$$

$$\frac{2n}{2} = \frac{-4}{2} \qquad \text{Division Property of Equality}$$

$$n = \boxed{} \qquad \text{Simplify.}$$

So, the solution to the equation is $n = -2$.

Check the solution.

$$6n - 1 = 4n - 5 \qquad \text{Write the equation.}$$

$$6(\boxed{}) - 1 = 4(\boxed{}) - 5 \qquad \text{Replace } n \text{ with } -2.$$

$$-12 - 1 = -8 - 5 \qquad \text{Multiply.}$$

$$\boxed{} = \boxed{} \qquad \text{Simplify. The solution, } -2, \text{ is correct.}$$

Check

Solve $10 - 3x = -5 + 2x$.

Show your work here

Talk About It!

Describe how you can use algebra tiles to solve the equation. Does it matter that the variable is n?

Go Online You can complete an Extra Example online.

Example 2 Solve Equations with Rational Coefficients

Solve $\frac{2}{3}x - 1 = 9 - \frac{1}{6}x$. **Check your solution.**

Method 1 Solve the equation using fractions.

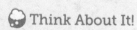
$\frac{2}{3}x - 1 = 9 - \frac{1}{6}x$	Write the equation.
$\frac{\Box}{6}x - 1 = 9 - \frac{\Box}{6}x$	The common denominator of the coefficients is 6.
$+\frac{1}{6}x \quad = \quad +\frac{1}{6}x$	Addition Property of Equality
$\frac{\Box}{\Box}x - 1 = \Box$	Simplify.
$+1 = +1$	Addition Property of Equality
$\frac{5}{6}x \quad = \Box$	Simplify.
$\left(\frac{6}{5}\right)\frac{5}{6}x = 10\left(\frac{6}{5}\right)$	Multiplication Property of Equality
$x = \Box$	Simplify.

Method 2 Solve the equation by using the LCD to eliminate the fractions.

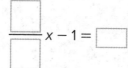
$\frac{2}{3}x - 1 = 9 - \frac{1}{6}x$	Write the equation.
$\frac{4}{6}x - 1 = 9 - \frac{1}{6}x$	Rewrite with common denominators.
$6\left(\frac{4}{6}x - 1\right) = \left(9 - \frac{1}{6}x\right)6$	Multiply by 6 to eliminate fractions.
$\Box - \Box = \Box - \Box$	Distributive Property
$+x \quad = \quad +x$	Addition Property of Equality
$\Box - 6 = 54$	Simplify.
$+6 = +6$	Addition Property of Equality
$5x \quad = \Box$	Simplify.
$\frac{5x}{5} = \frac{60}{5}$	Division Property of Equality
$x = \Box$	Simplify.

So, using either method, the solution to the equation is $x = 12$.

(continued on next page)

Check the solution.

$$\frac{2}{3}x - 1 = 9 - \frac{1}{6}x$$ Write the equation.

$$\frac{2}{3} \cdot \boxed{} - 1 = 9 - \frac{1}{6} \cdot \boxed{}$$ Replace x with 12.

$$\boxed{} - 1 = 9 - \boxed{}$$ Multiply.

$$7 = 7$$ Simplify. The solution, 12, is correct.

Check

Solve $\frac{2}{3}x + 5 = \frac{2}{5}x - 3$.

 Go Online You can complete an Extra Example online.

Pause and Reflect

Explain why you should use the original equation when checking the solution of an equation.

Record your observations here

Math History Minute

Alicia Dickenstein (1955-) is an Argentine mathematician and professor at the University of Buenos Aires. In 2015, she received the TWAS (The World Academy of Sciences) Prize for mathematics, which is awarded to individuals from developing countries who make outstanding contributions to science or mathematics. In 2018, Dickenstein was invited to speak at the World Meeting for Women in Mathematics in Rio de Janeiro, Brazil.

Example 3 Solve Equations with Rational Coefficients

Solve $2.3x + 2.8 = -1.2x + 9.8$. Check your solution.

$2.3x + 2.8 = -1.2x + 9.8$	Write the equation.
$\underline{+\,1.2x \qquad\quad = \boxed{}}$	Addition Property of Equality
$\boxed{} + 2.8 = \qquad 9.8$	Simplify.
$\underline{\quad\; -\,2.8 = \qquad -2.8}$	Subtraction Property of Equality
$3.5x \qquad\quad = \boxed{}$	Simplify.
$\dfrac{3.5x}{3.5} = \dfrac{7}{3.5}$	Division Property of Equality
$x = \boxed{}$	Simplify.

So, the solution to the equation is $x = 2$.

Check the solution.

$2.3x + 2.8 = -1.2x + 9.8$	Write the equation.
$2.3(2) + 2.8 = -1.2(2) + 9.8$	Replace x with 2.
$4.6 + 2.8 = -2.4 + 9.8$	Multiply.
$7.4 = 7.4$	Simplify. The solution, 2, is correct.

Check

Solve $4.06x + 3.22 = 27.16 - 1.26x$.

Show your work here

Go Online You can complete an Extra Example online.

Practice

⟲ **Go Online** You can complete your homework online.

Solve each equation. Check your solution. (Examples 1–3)

1. $-2a - 9 = 6a + 15$

2. $14 + 3n = 5n - 6$

3. $\frac{1}{2}x - 5 = 10 - \frac{3}{4}x$

4. $\frac{2}{3}y + 1 = \frac{1}{6}y + 8$

5. $5.4p + 13.1 = -2.6p + 3.5$

6. $0.15w + 0.35 = 0.22w - 0.14$

Test Practice

7. Twelve more than seven times a number equals the number less six. Solve the equation $7x + 12 = x - 6$ to find the number, x.

8. Equation Editor Solve the equation shown for x.

$$3x - 15 = 17 - x$$

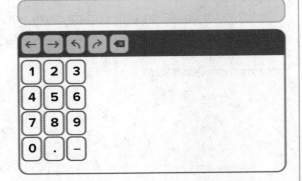

Lesson 3-1 • Solve Equations with Variables on Each Side **135**

Apply

9. A plumbing company charges $35 per hour plus a $25 travel charge for a service call. Another plumbing company charges $40 per hour for a service call with no travel charge. Solve the equation $35h + 25 = 40h$ to find how many hours h a service call must be for the two companies to charge the same amount.

10. For an annual membership fee of $186, Jacy can join the local pool that would allow him to take diving lessons for $10 each session with an instructor. Without the membership, the pool charges $25.50 for each diving lesson with an instructor. Solve the equation $186 + 10x = 25.5x$ to find how many diving lessons x Jacy can take for the cost to be the same with and without a membership.

11. **MP** **Identify Structure** Explain how the Distributive Property can be used to eliminate the fractions in the equation $\frac{1}{5}x + 8 = \frac{1}{10}x + 9$.

12. **MP** **Find the Error** A student solved the equation $6n + 8 = 4n - 9$. Find his mistake and correct it.

$$6n + 8 = 4n - 9$$
$$6n + 8 - 8 = 4n - 9 - 8$$
$$6n = 4n - 1$$
$$6n - 4n = 4n - 4n - 1$$
$$2n = -1$$
$$n = -0.5$$

13. **MP** **Justify Conclusions** Determine if the statement is *true* or *false*. Justify your response.

    To solve an equation with variables on each side, such as $-9 + 3x = 8x + 6$, you always first need to add or subtract the variable terms from each side.

14. Write an equation with variables on each side of the equals sign that has a solution of -2.

Write and Solve Equations with Variables on Each Side

I Can... write linear equations in one variable with rational coefficients and use the properties of equality to solve them.

Explore Write and Solve Equations with Variables on Each Side

Online Activity You will explore how to write an equation with variables on each side to solve a real-world problem.

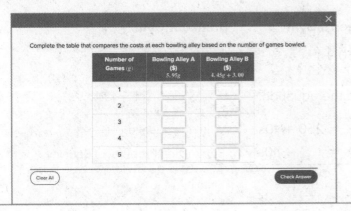

Complete the table that compares the costs at each bowling alley based on the number of games bowled.

Number of Games (g)	Bowling Alley A ($) $5.95g$	Bowling Alley B ($) $4.45g + 3.00$
1		
2		
3		
4		
5		

Clear All   Check Answer

Learn Write and Solve Equations with Variables on Each Side

You can represent many real-world problems using equations.

A music streaming website offers two plans. The first plan costs $0.99 per song plus an initial fee of $25. The second plan costs $1.50 per song plus an initial fee of $10. For how many songs will the two plans cost the same?

Words
a fee of $25 plus $0.99 per song is the same as a fee of $10 plus $1.50 per song
Variables
Let s represent the number of songs.
Equation
$25 + 0.99s = 10 + 1.50s$

Talk About It!

What will a solution to the equation represent within the context of the problem?

Think About It!

What is the unknown in this problem?

 Talk About It!

Suppose a customer plans to have a personal training session once a month, for one year. Which gym is more cost effective for them to choose? Explain.

 Example 1 Write and Solve Equations with Variables on Each Side

Green's Gym charges a one-time application fee of $50 plus $30 per session for a personal trainer. Breakout Gym charges an annual fee of $250 plus $10 for each session with a trainer.

For how many sessions is the cost of the two plans the same? Write and solve an equation to represent this problem. Check your solution.

Part A Write an equation to represent the problem.

Let s represent the number of _____. Write an equation that models when the cost of the two plans are equal to each other.

Green's Gym	Breakout Gym
50 + ☐ =	☐ + 10s

Part B Solve the equation.

$50 + 30s = 250 + 10s$	Write the equation.
$\underline{-10s = -10s}$	Subtraction Property of Equality
$50 + \boxed{} = \boxed{}$	Simplify.
$\underline{-50 = -50}$	Subtraction Property of Equality
$\boxed{} = \boxed{}$	Simplify.
$\dfrac{20s}{20} = \dfrac{200}{20}$	Division Property of Equality
$s = \boxed{}$	Simplify.

So, the cost is the same for 10 personal trainer sessions.

Check the solution.

Green's Gym		Breakout Gym
$50 + 30s = 50 + 30(10)$	Replace s with 10.	$250 + 10s = 250 + 10(10)$
$= 50 + 300$	Multiply.	$= 250 + 100$
$= \boxed{}$	Simplify.	$= \boxed{}$

Check

A container has 130 gallons of water and is being filled at a rate of $\frac{1}{4}$ gallon each second. Another container has 200 gallons of water and is draining at a rate of $\frac{1}{3}$ gallon each second.

Part A

Which equation could be used to determine s, the number of seconds, when the two containers have the same amount of water?

(A) $130 + \frac{1}{3}s = 200 - \frac{1}{4}s$

(B) $130 + \frac{1}{4}s = 200 - \frac{1}{3}s$

(C) $200 + \frac{1}{3}s = 130 - \frac{1}{4}s$

(D) $4 + \frac{1}{130}s = 3 - \frac{1}{200}s$

Part B

After how many seconds will the two containers have the same amount of water?

Show your work here

 Go Online You can complete an Extra Example online.

Example 2 Write and Solve Equations with Variables on Each Side

Ryan's Rentals charges $40 per day plus $0.25 per mile. Road Trips charges $25 per day plus $0.45 per mile.

For what number of miles is the daily cost of renting a car the same? Write and solve an equation to represent this problem. Check your solution.

Part A Write an equation.

Let m represent the number of _____. Write an equation that models when the cost of the two rentals are equal.

Ryan's Rentals		**Road Trips**
$40 + \boxed{}$	$=$	$\boxed{} + 0.45m$

(continued on next page)

Part B Solve the equation.

$40 + 0.25m = 25 + 0.45m$	Write the equation.
$\underline{ -0.25m = -0.25m}$	Subtraction Property of Equality
$\boxed{} = 25 + \boxed{}$	Simplify.
$\underline{-25 = -25}$	Subtraction Property of Equality
$\boxed{} = \boxed{}$	Simplify.
$\dfrac{15}{0.20} = \dfrac{0.20m}{0.20}$	Division Property of Equality
$\boxed{} = m$	Simplify.

So, the cost is the same for 75 miles in one day.

Check the solution.

Ryan's Rentals		**Road Trips**
$40 + 0.25m$		$25 + 0.45m$
$= 40 + (0.25)75$	Replace m with 75.	$= 25 + (0.45)75$
$= 40 + \boxed{}$	Multiply.	$= 25 + \boxed{}$
$= \boxed{}$	Simplify.	$= \boxed{}$

Check

Annie is comparing the cost to ship a package. One shipping company charges $7 for the first pound and $0.20 for each additional pound a package weighs. Another shipping company charges $5 for the first pound and $0.30 for each additional pound.

Part A

Which equation could be used to determine p, the number of pounds, when the costs for the two shipping companies are the same?

Ⓐ $7 + 0.30p = 5 + 0.20p$ Ⓒ $7 + 0.20p = 5 + 0.30p$

Ⓑ $0.20 + 7p = 0.30 + 5p$ Ⓓ $7 + 20p = 5 + 30p$

Part B

For how many pounds is the cost of the two shipping companies the same?

Show your work here

🔎 **Go Online** You can complete an Extra Example online.

💬 **Talk About It!**

Which company is more cost effective to use if you plan to drive 50 miles in one day? 150 miles? Explain.

🌐 Apply Home Improvement

Suppose you are replacing the carpet in a living room where the length of the living room is five feet shorter than twice its width, *w*. Tack strip is placed around the perimeter of the room, which is equal to five times the width. If carpet costs $2.99 a square foot, what is the total cost to carpet the living room?

1 What is the task?

Make sure you understand exactly what question to answer or problem to solve. You may want to read the problem three times. Discuss these questions with a partner.

First Time Describe the context of the problem, in your own words.
Second Time What mathematics do you see in the problem?
Third Time What are you wondering about?

2 How can you approach the task? What strategies can you use?

3 What is your solution?

Use your strategy to solve the problem.

4 How can you show your solution is reasonable?

✏️ **Write About It!** Write an argument that can be used to defend your solution.

📡 Go Online
Watch the animation.

w ft

(2*w* − 5) ft

💬 Talk About It!
How could you solve this problem another way?

Check

A rectangular bathroom with the side lengths shown is being covered with tiles, where x is the length, in feet, of a square tile. The perimeter is equal to $48x - 6$. If each square foot of tile costs \$8.49, what is the total cost to tile the bathroom?

$6x$

$12x$

Show your work here

🔄 **Go Online** You can complete an Extra Example online.

Pause and Reflect

How will you study the concepts in today's lesson? Describe some steps you can take.

Record your observations here

Practice

Go Online You can complete your homework online.

Write and solve an equation for each exercise. Check your solution.
(Examples 1 and 2)

1. Marko has 45 comic books in his collection, and Tamara has 61 comic books. Marko buys 4 new comic books each month and Tamara buys 2 comic books each month. After how many months will Marko and Tamara have the same number of comic books?

2. A fish tank has 150 gallons of water and is being drained at a rate of $\frac{1}{2}$ gallon each second. A second fish tank has 120 gallons of water and is being filled at a rate of $\frac{1}{4}$ gallon each second. After how many seconds will the two fish tanks have the same amount of water?

3. Shipping Company A charges $14 plus $2.25 a pound to ship overnight packages. Shipping Company B charges $20 plus $1.50 a pound to ship an overnight package. For what weight is the charge the same for the two companies?

4. A bicycle rental company charges a $20 fee plus $5.50 per hour to rent a bicycle. Another bicycle rental company charges a $15 fee plus $6.50 per hour to rent a bicycle. For what number of hours is the cost for the rental the same?

Test Practice

5. **Open Response** Deanna and Lulu are playing games at the arcade. Deanna starts with $15, and the machine she is playing costs $0.75 per game. Lulu starts with $13, and her machine costs $0.50 per game. After how many games will the two friends have the same amount of money remaining? Let *g* represent the number of games.

Equation: _____

Number of Games: _____

Apply

6. Aiden is replacing the tile in a rectangular kitchen. The length of the kitchen is nine feet shorter than three times its width, w. The perimeter of the kitchen is six times the width. If tiles cost $1.69 a square foot, what is the total cost to tile the kitchen?

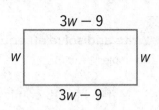

7. Hailey is putting up a fence in the shape of an isosceles triangle in her backyard. The fence has side lengths as shown, where x represents the number of feet in each fence section. The perimeter of the fence can be covered using 8 total fence sections represented by the expression $8x$. If fencing costs $6.50 a foot, what would be the total cost of the fence?

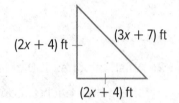

8. **Create** Write a real-world problem that can be solved using the equation $9 + 4x = 21 - 2x$.

9. Ling worked three more hours on Tuesday than she did on Monday. On Wednesday, she worked one hour more than twice the number of hours that she worked on Monday. The total number of hours is two more than five times the number of hours worked on Monday. Write and solve an equation to find the number of hours she worked on Monday.

10. MP **Find the Error** A student wrote the equation $22 + 4 = 6s + 12s$ to represent the problem shown at the right. Find his mistake and correct it.

Darnell and Emma are college students. Darnell currently has 22 credits and he plans on taking 6 credits per semester. Emma has 4 credits and plans to take 12 credits per semester. After how many semesters, s, will Darnell and Emma have the same number of credits?

Solve Multi-Step Equations

I Can... solve multi-step linear equations with rational coefficients by using the Distributive Property and combining like terms.

What Vocabulary Will You Learn?
like terms

Learn Solve Multi-Step Equations

Some equations contain expressions with grouping symbols on one or both sides of the equals sign.

To solve equations like this, first expand the expressions that contain grouping symbols. Then solve the equation, combining any **like terms** and using the Properties of Equality.

Go Online Watch the animation to see how to solve the multi-step equation $-5(2x + 3) - x = 4(x + 11) + 1$.

The animation shows that you can use the Distributive Property and the Properties of Equality to solve a multi-step equation that contains expressions with grouping symbols.

$-5(2x + 3) - x = 4(x + 11) + 1$	Write the equation.
$-10x - 15 - x = 4x + 44 + 1$	Expand the expressions using the Distributive Property.
$-11x - 15 = \quad 4x + 45$	Combine the like terms $-10x$ and $-x$. Combine the like terms 44 and 1.
$\underline{+ 11x \qquad = + 11x}$	Addition Property of Equality
$-15 = 15x + 45$	Simplify.
$\underline{-45 = \qquad - 45}$	Subtraction Property of Equality.
$-60 = 15x$	Simplify.
$\dfrac{-60}{15} = \dfrac{15x}{15}$	Division Property of Equality
$-4 = x$	Simplify.

The solution of the equation is $x = $ _____.

Talk About It!

How can you make sure that you solved the equation correctly?

Example 1 Solve Multi-Step Equations

Solve $3(8x + 12) - 15x = 2(3 - 3x)$. Check your solution.

$$3(8x + 12) - 15x = 2(3 - 3x)$$ Write the equation.

$$\boxed{} + \boxed{} - 15x = \boxed{} - \boxed{}$$ Distributive Property

$$\boxed{} + 36 = 6 - 6x$$ Combine like terms.

$$\underline{+ 6x \qquad\quad = \quad + 6x}$$ Addition Property of Equality

$$\boxed{} + 36 = 6$$ Simplify.

$$\underline{- 36 = - 36}$$ Subtraction Property of Equality

$$15x = \boxed{}$$ Simplify.

$$\frac{15x}{\boxed{}} = \frac{-30}{\boxed{}}$$ Division Property of Equality

$$x = \boxed{}$$ Simplify.

So, the solution to the equation is $x = -2$.

Check the solution.

$$3(8x + 12) - 15x = 2(3 - 3x)$$ Write the equation.

$$3\left[8\left(\boxed{}\right) + 12\right] - 15\left(\boxed{}\right) = 2\left[3 - 3\left(\boxed{}\right)\right]$$ Replace x with -2.

$$-12 - (-30) = 2(9)$$ Multiply.

$$\boxed{} = \boxed{}$$ Simplify.

Check

Solve $8(-2 + x) - 3x = 6x + 13$.

Show your work here

Go Online You can complete an Extra Example online.

Example 2 Solve Multi-Step Equations

Solve $0.3(10 - 5x) = 31.5 - (8x + 9)$. Check your solution.

$0.3(10 - 5x) = 31.5 - (8x + 9)$	Write the equation.
$\boxed{} - \boxed{} = 31.5 - (8x + 9)$	Distributive Property
$3 - 1.5x = 31.5 + (-8x - 9)$	Rewrite using the additive inverse.
$3 - 1.5x = \boxed{} - 8x$	Combine like terms.
$\underline{-3 \qquad\quad = -3}$	Subtraction Property of Equality
$\boxed{} = \boxed{} - 8x$	Simplify.
$\underline{+\ 8x = \qquad\quad +\ 8x}$	Addition Property of Equality
$\boxed{} = 19.5$	Simplify.
$\dfrac{6.5x}{\boxed{}} = \dfrac{19.5}{\boxed{}}$	Division Property of Equality
$x = \boxed{}$	Simplify.

So, the solution to the equation is $x = 3$.

Check the solution.

To check your solution, replace x with 3 in the original equation.

$0.3(10 - 5x) = 31.5 - (8x + 9)$	Write the equation.
$0.3\left(10 - 5 \cdot \boxed{}\right) = 31.5 - \left(8 \cdot \boxed{} + 9\right)$	Substitute.
$0.3\left(10 - \boxed{}\right) = 31.5 - \left(\boxed{} + 9\right)$	Multiply.
$0.3\left(\boxed{}\right) = 31.5 - \boxed{}$	Simplify.
$\boxed{} = \boxed{}$	Simplify.

Check

Solve $15 - (7x + 20) = 0.5(6x - 14)$.

Show your work here

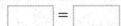 **Go Online** You can complete an Extra Example online.

Talk About It!

How does the Additive Inverse Property allow the parentheses to be removed in order to simplify the right side of the equation?

Example 3 Solve Multi-Step Equations

Solve $\frac{1}{2}(6x - 4) + 6x = -12\left(\frac{1}{6}x + 2\right)$. Check your solution.

$$\frac{1}{2}(6x - 4) + 6x = -12\left(\frac{1}{6}x + 2\right)$$ Write the equation.

$\boxed{} - \boxed{} + 6x = \boxed{} - \boxed{}$ Distributive Property

$\boxed{} - 2 \; = \; -2x - 24$ Combine like terms.

$\dfrac{+ 2x \qquad\quad = + 2x}{}$ Addition Property of Equality

$\boxed{} - 2 \; = -24$ Simplify.

$\dfrac{+ 2 \; = + 2}{}$ Addition Property of Equality

$11x = \boxed{}$ Simplify.

$\dfrac{11x}{\boxed{}} = \dfrac{-22}{\boxed{}}$ Division Property of Equality

$x = \boxed{}$ Simplify.

So, the solution to the equation is $x = -2$.

Check

Solve $\frac{3}{4}(-5x - 8) = -6(x + 4) + \frac{1}{4}x$.

Show your work here

🔄 **Go Online** You can complete an Extra Example online.

📓 **Foldables** It's time to update your Foldable, located in the Module Review, based on what you learned in this lesson. If you haven't already assembled your Foldable, you can find the instructions on page FL1.

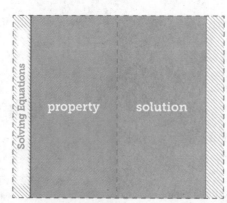

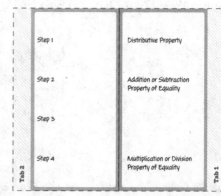

Step 1   Distributive Property

Step 2   Addition or Subtraction Property of Equality

Step 3

Step 4   Multiplication or Division Property of Equality

Practice

Go Online You can complete your homework online.

Solve each equation. Check your solution. (Examples 1–3)

1. $-g + 2(3 + g) = -4(g + 1)$

2. $-8 - x = -3(2x - 4) + 3x$

3. $0.6(4 - 2x) = 20.5 - (3x + 10)$

4. $12 - (4y + 8) = 0.5(8y - 16)$

5. $\frac{1}{2}(-4 + 6n) = \frac{1}{3}n + \frac{2}{3}(n + 9)$

6. $\frac{1}{5}(5x - 5) + 3x = -9\left(\frac{1}{3}x + 4\right)$

Test Practice

7. Equation Editor Solve the equation shown for q.

$$2\left(\frac{1}{2}q + 1\right) = -3(2q - 1) + 8q + 4$$

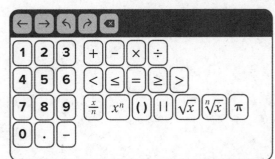

Apply

8. At a health club, 96 members participate in the morning workout, and this number has been increasing by 2 people per week. In the afternoon workout, there are 80 members, and this number has been decreasing by 3 people per week. Solve the equation $96 + 2w = 2(80 - 3w)$ to find in how many weeks w the number of people working out in the morning will be double the number of people working out in the afternoon.

9. The triangle and the square shown have the same perimeter. Solve the equation $3x + 4x + 5x = 4(x + 2)$ to find the value of x. Then find the length of one side of the square.

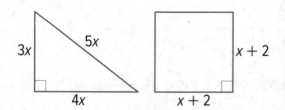

10. 🔵 **Find the Error** A student solved the equation $3(-4 + x) - 5x = 7x + 15$. Find her mistake and correct it.

$$3(-4 + x) - 5x = 7x + 15$$
$$-12 + 3x - 15x = 7x + 15$$
$$-12 - 12x = 7x + 15$$
$$-12 - 12x + 12x = 7x + 12x + 15$$
$$-12 = 19x + 15$$
$$-12 - 15 = 19x + 15 - 15$$
$$-27 = 19x$$
$$-1\frac{8}{19} = x$$

11. 🔵 **Identify Structure** Describe the role of the Distributive Property when solving multi-step equations that contain expressions with grouping symbols.

12. Suppose your friend says he can solve the equation $15 - (2x + 6) = \frac{1}{2}(7x - 4)$ by first using the Multiplication Property of Equality to multiply each side of the equation by 2. Is he correct? Justify your response.

Write and Solve Multi-Step Equations

I Can... write and solve multi-step linear equations with rational coefficients by using the Distributive Property and combining like terms.

Explore Translate Problems into Equations

Online Activity You will explore how to model a real-world problem with a multi-step equation with variables on both sides.

Complete the table to show the cost, in dollars, of shirts for the number of players shown.

Number of Players	Cost of Shirt with Name, $
1	$20 + n$
2	$2(20 + n)$
3	$3(20 + n)$
4	
5	
6	

Learn Write and Solve Multi-Step Equations

You can represent many real-world problems using multi-step equations.

Four friends bought zoo tickets and spent $9.50 each on wristbands for the zoo rides. The same day, a group of 5 different friends bought zoo tickets and spent $3 each on the sting ray exhibit. The group of 5 friends also rented a locker for $8. If both groups spent the same amount, how much did each person pay for a zoo ticket?

Words
4 times the cost of a ticket and a $9.50 wrist band is the same as an $8 locker plus five times the cost of a ticket and a $3 exhibit

Variables
Let t represent the cost of each ticket.

Equation
$4(t + 9.50) = 8 + 5(t + 3)$

Talk About It!
Describe another real-world situation in which you might need to solve a multi-step equation in order to solve a problem.

Mrs. Hill is designing a rectangular vegetable garden for her backyard. The width of the garden is $11\frac{1}{2}$ feet shorter than twice its length.

If the perimeter of the garden is 37 feet, what is the length of the garden? Write and solve an equation. Check your solution.

Part A Write an equation.
Let ℓ represent the _____, in feet, of the garden. The width of the garden is twice the length _____ $11\frac{1}{2}$ feet. So, the expression $2\ell - 11\frac{1}{2}$ represents the _____ of the garden.

Write an equation that models the perimeter.

$$P = 2\ell + 2w$$

$$\boxed{} = 2\ell + 2\left(\boxed{}\right)$$

Part B Solve the equation.

$37 = 2\ell + 2\left(2\ell - 11\frac{1}{2}\right)$	Write the equation.
$37 = 2\ell + \boxed{} - \boxed{}$	Distributive Property
$37 = \boxed{} - 23$	Combine like terms.
$\underline{+23 = + 23}$	Addition Property of Equality
$\boxed{} = 6\ell$	Simplify.
$\dfrac{60}{6} = \dfrac{6\ell}{6}$	Division Property of Equality
$\boxed{} = \ell$	Simplify.

So, the length of the garden is 10 feet.

Check the solution.

Length: $\boxed{}$ feet

Width: $2\ell - 11\frac{1}{2} = 2(10) - 11\frac{1}{2}$ or $\boxed{}$ feet

Perimeter: $2\ell - 2w = 2(10) + 2(8.5)$ or $\boxed{}$ feet

Think About It!
What is the unknown in the problem?

Talk About It!
What is another equation you can write to represent the perimeter of the garden, other than $P = 2\ell + 2w$?

Check

The hallway in Oscar's house is in the shape of a rectangle. The length of the hallway is $\frac{1}{2}$ foot longer than twice its width. The perimeter of the hallway is 31 feet.

Part A

Which equation(s) can be used to find the width w of the hallway? Select all that apply.

☐ $31 = \frac{1}{2} + 4w$ ☐ $31 = 1 + 6w$ ☐ $31 = 2 + 2w$

☐ $31 = 2\left(\frac{1}{2} + 2w\right) + 2w$ ☐ $31 = \frac{1}{2} + 2w$ ☐ $31 = 2\left(\frac{1}{2} + 3w\right)$

Part B

What is the width of the hallway?

Show your work here

🔵 **Go Online** You can complete an Extra Example online.

🌐 Example 2 Write and Solve Multi-Step Equations

Mr. Murphy's class of 20 students is going on a field trip to the science center. They will also watch the 3-D movie. Mrs. Todd's class of 15 students is going on a field trip to the art museum and will take the audio tour. Admission to the art museum is 2.5 times that of the science center's entry fee, as shown in the table.

If the total cost is the same at both the science center and the art museum, what is the entry fee per student to the science center? Check your solution.

Science Center	Art Museum
Entry fee: $x per student	Admission: $2.5x per student
3-D movie: $2.50 per student	Audio Tour: $1 per student

Part A Write an equation.

Let x represent the cost of the entry fee per student to the science center. The cost of the trip to the science center can be represented by the expression $(x + 2.50)$ for each student to pay the entry fee and 3-D movie. The cost of the trip to the Art Museum can be represented by the expression $(2.5x + 1)$ for admission and the audio tour for each student. Complete the equation that models the total cost.

$$\boxed{}(x + 2.50) = \boxed{}(2.5x + 1)$$

(continued on next page)

> 😎 **Think About It!**
>
> What quantity will the variable represent?

Can you solve the equation without using the Distributive Property? Why is the Distributive Property helpful?

Part B Solve the equation.

$20(x + 2.50) = 15(2.5x + 1)$ Write the equation.

$\boxed{} + \boxed{} = \boxed{} + \boxed{}$ Distributive Property

$\underline{-20x \qquad\qquad = -20x}$ Subtraction Property of Equality

$50 = \boxed{} + 15$ Simplify.

$\underline{-15 = \qquad\qquad -15}$ Subtraction Property of Equality

$\boxed{} = 17.5x$ Simplify.

$\dfrac{35}{\boxed{}} = \dfrac{17.5x}{\boxed{}}$ Division Property of Equality

$\boxed{} = x$ Simplify.

So, the entry fee per student to the science center is $2.

Check the solution.

Science Center: $20(x + 2.5) = 20(2 + 2.50)$ or $\$\boxed{}$

Art Museum: $15(2.5x + 1) = 15(2.5(2) + 1)$ or $\$\boxed{}$

Check

A group of 6 friends went to a basketball tournament game. They each bought a ticket for x dollars and spent $8.50 each on snacks. Another group of 4 friends went to the championship game and paid twice as much for each of their tickets as the first group. The group of 4 friends also spent $6.50 each on snacks.

Part A

Which equation can be used to find the cost of each ticket for the group of 6 friends?

Ⓐ $8.50(x + 6) = 4(2x + 6.50)$

Ⓑ $6(x + 8.50) = 2(4x + 6.50)$

Ⓒ $8.50(x + 6) = 6.50(2x + 4)$

Ⓓ $6(x + 8.50) = 4(2x + 6.50)$

Part B

What is the cost of each ticket for the group of 6 friends?

Show your work here

🇰 **Go Online** You can complete an Extra Example online.

🌐 Apply Business Finance

The table shows the hours worked by employees at a coffee shop last month. If each employee earns $15 per hour and the total number of hours worked is represented by $7m + 19$ where m represents the number of hours Mai worked.

Employee	Hours Worked
Shantel	48
Lorenzo	$2m + 7$
Jamie	$3.5(m - 6)$
Mai	m

What was the total payroll, or amount paid to the employees?

1 What is the task?

Make sure you understand exactly what question to answer or problem to solve. You may want to read the problem three times. Discuss these questions with a partner.

First Time Describe the context of the problem, in your own words.
Second Time What mathematics do you see in the problem?
Third Time What are you wondering about?

2 How can you approach the task? What strategies can you use?

3 What is your solution?

Use your strategy to solve the problem.

4 How can you show your solution is reasonable?

🖊 **Write About It!** Write an argument that can be used to defend your solution.

Check

The table shows the number of minutes Declan participated in various activities last week. The total number of minutes he participated in all of the activities was $5x + 15.75$. What is the ratio of the number of minutes Declan rode his bike to the number of minutes he had soccer practice?

Activity	Time (minutes)
Ride Bike	x
Soccer	$2.5(x + 3) + 2.5$
Swim	15.5
Dog Walk	$3(x - 13.25)$

Go Online You can complete an Extra Example online.

Pause and Reflect

Write one sentence about today's lesson for each of the categories: Who, What, Where, How, and Why.

Practice

Write and solve an equation for each exercise. Check your solution.
(Examples 1 and 2)

1. Mr. Reed is drawing a blueprint of a rectangular patio. The width of the patio is $40\frac{3}{4}$ feet shorter than twice its length. The perimeter of the patio is $86\frac{1}{2}$ feet. What is the length of the patio?

2. The Yearbook Club is going to an amusement park, and each of their 12 members will pay for admission and will also help pay for parking. The Robotics Club is going to a waterpark, and each of their 14 members will pay for admission and will also purchase a meal ticket. Admission to the amusement park is 1.5 times that of the waterpark's admission, as shown in the table. If the total cost is the same at both the amusement park and the waterpark, what is the admission per student to the waterpark?

Amusement Park	Waterpark
Admission: $1.5x$ per student	Admission: x per student
Parking: $2 per student	Meal Ticket: $10.50 per student

Test Practice

3. **Open Response** Arjun purchased 5 tickets to a play, each with the same price. He was also charged an online service fee of $3.50 per ticket. Emilia purchased 3 tickets to the same play and paid twice as much for her tickets as Arjun. Emilia was also charged a service fee of $2.75 per ticket. If they spent the same amount, what is the cost of each of the tickets Arjun purchased? Let t represent the cost of each of Arjun's tickets.

Equation:

Cost of a Ticket:

Apply

4. Four siblings have a dog walking business. The table shows the hours worked by each sibling. Each sibling earns $25.50 per hour and the total number of hours worked is represented by $10h + 15$, where h represents the number of hours Michael worked. What was the total amount the siblings earned?

Sibling	Hours Worked
Martin	$2.5h + 3$
Emilio	$4(h - 2)$
Michael	h
Mario	31

5. **Create** Write a real-world problem that can be solved using a multi-step equation. Then write and solve an equation for your problem.

6. (MP) **Persevere with Problems** Elijah put $2x + 3$ dollars in the bank the first week. The following week he doubled the first week's savings and put that amount in the bank. The next week, he doubled what was in the bank and put that amount in the bank. He now has $477 in the bank. Write and solve an equation to find how much money he put in the bank the first week.

7. (MP) **Find the Error** A student wrote the equation $4(x + 5.5) = 3(1.5x + 4.75)$ to represent the problem shown. Find her mistake and correct it.

Petra and her 4 friends went to the movies. They each bought a ticket for x dollars and spent $5.50 each on snacks. Valentina and her 3 friends went to the movies and paid 1.5 times as much for each of their tickets as Petra. Everyone in Valentina's group also spent $4.75 each on snacks. If both groups spent the same amount, what is the cost of a ticket for Petra's group?

Determine the Number of Solutions

I Can... identify the number of solutions of a linear equation in one variable by simplifying each side and comparing coefficients and constants.

Explore Number of Solutions

Online Activity You will use Web Sketchpad to explore equations with one solution, no solution, and infinitely many solutions.

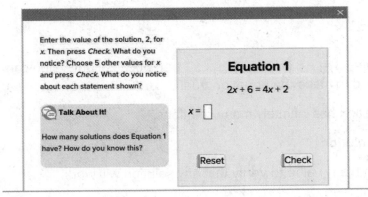

Enter the value of the solution, 2, for x. Then press *Check*. What do you notice? Choose 5 other values for x and press *Check*. What do you notice about each statement shown?

Talk About It!

How many solutions does Equation 1 have? How do you know this?

Equation 1

$2x + 6 = 4x + 2$

$x = \boxed{}$

Reset   Check

Learn Number of Solutions

The solution to an equation is the value of the variable that makes the equation true. Some equations have one solution, while some equations have no solution. When this occurs, no value will make the equation true. Other equations may have infinitely many solutions. When this occurs, the equation is true for every value of the variable.

	No Solution	One Solution	Infinite Solutions
Symbols	$a = b$	$x = a$	$a = a$
Examples	$3x + 4 = 3x$ $4 = 0$	$2x = 20$ $x = 10$	$4x + 2 = 4x + 2$ $2 = 2$
	Since $4 \neq 0$, there is no solution.	Since $x = 10$, there is one solution.	Since $2 = 2$, all of the values of x are solutions.

Talk About It!

Study the structure of the equations with no solution compared to those with infinitely many solutions. What do you notice?

😮 **Think About It!**

How will you know if the equation has one solution, no solution, or infinitely many solutions?

Example 1 Equations with Infinitely Many Solutions

Solve $6(x - 3) + 10 = 2(3x - 4)$. Determine whether the equation has one solution, no solution, or infinitely many solutions. Check your solution.

$$6(x - 3) + 10 = 2(3x - 4)$$ Write the equation.

$$\boxed{} - \boxed{} + 10 = \boxed{} - \boxed{}$$ Distributive Property

$$6x - \boxed{} = 6x - 8$$ Combine like terms.

$$\underline{-6x \qquad\qquad = -6x}$$ Subtract 6x from each side.

$$-8 = -8$$ Simplify.

The equation $-8 = -8$ is _____ true because any value can be substituted to make the equation true.

So, the equation has infinitely many solutions.

Check the solution.

Replace x with any value to verify that any solution will work.

$$6(x - 3) + 10 = 2(3x - 4)$$ Write the equation.

$$6\left(\boxed{} - 3\right) + 10 = 2\left(3 \cdot \boxed{} - 4\right)$$ Replace x with 4.

$$16 = 16$$ Simplify.

💬 **Talk About It!**

After simplifying each side of the equation, the equation becomes $6x - 8 = 6x - 8$. Why is it not necessary to finish solving to find the number of solutions?

Check

Which equation has an infinite number of solutions?

Ⓐ $2(3c - 6) - 2c = 4(c + 4)$

Ⓑ $3(2p - 1) = 2(p + 10) + 1$

Ⓒ $8(x - 9) = 6(2x - 12) - 4x$

Ⓓ $5(x - 2) - 20 = -5(x - 6)$

Show your work here

🅑 **Go Online** You can complete an Extra Example online.

Example 2 Equations with No Solution

Solve $8(4 - 2x) = 4(3 - 5x) + 4x$. Determine whether the equation has one solution, no solution, or infinitely many solutions.

$8(4 - 2x) = 4(3 - 5x) + 4x$ Write the equation.

$\boxed{} - \boxed{} = \boxed{} - \boxed{} + 4x$ Distributive Property

$32 - 16x = 12 - \boxed{}$ Combine like terms.

$\underline{+ 16x = + 16x }$ Addition Property of Equality

$\boxed{} = 12$ Simplify.

The equation $32 = 12$ is _____ true because no values can be substituted to make the equation true.

So, the equation has no solution.

Check

Which equation has no solution?

Ⓐ $5(b + 3) - 2b = 2b + 3$

Ⓑ $4(d - 3) + 5 = 3(d + 2) - 7$

Ⓒ $3(x + 5) = 5(x + 3) - 2x$

Ⓓ $-10y + 18 = -3(5y - 7) + 5y$

 Think About It!

What are some possible first steps to solving this equation?

 Talk About It!

After simplifying each side of the equation, the equation becomes $32 - 16x = 12 - 16x$. Without continuing to solve, how can you determine that the equation has no solution just by studying it?

Go Online You can complete an Extra Example online.

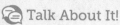

Learn Analyze Equations to Determine the Number of Solutions

It is possible to determine the number of solutions to an equation without actually solving it. The number of solutions to an equation can be found after each side has been simplified.

Complete the table to indicate whether the coefficients and constants are *the same* or *different*.

	No Solution	One Solution	Infinite Solutions
Equation	$6x + 3 = 6x + 1$	$6x + 3 = 4x + 1$	$6x + 3 = 6x + 3$
Coefficients	the same		the same
Constants		different or the same	

Example 3 Create Equations with Infinitely Many Solutions

What numbers would complete the equation so that it has infinitely many solutions?

$$6x - x + 4 + 2x = \boxed{?}\ x + \boxed{?}$$

$$6x - x + 4 + 2x = \boxed{?}\ x + \boxed{?} \qquad \text{Write the equation.}$$

$$\boxed{} + 4 = \boxed{?}\ x + \boxed{?} \qquad \text{Combine like terms.}$$

$$7x + 4 = \boxed{}\ x + \boxed{} \qquad \begin{array}{l}\text{The coefficients and constants}\\ \text{must be the same on each side.}\end{array}$$

So, $6x - x + 4 + 2x = 7x + 4$ is the equation with infinitely many solutions.

Check

Complete the equation with values that will result in an equation with infinitely many solutions.

$$4x - 2(x + 5) = \boxed{}\ x - \boxed{}$$

Show your work here

Go Online You can complete an Extra Example online.

Example 4 Create Equations with No Solution

What numbers would complete the equation so that it has no solution?

$3(2x + 4) - x = \boxed{?} \; x + \boxed{?}$

$3(2x + 4) - x = \boxed{?} \; x + \boxed{?}$ Write the equation.

$\boxed{} + \boxed{} - x = \boxed{?} \; x + \boxed{?}$ Distributive Property

$\boxed{} + 12 = \boxed{?} \; x + \boxed{?}$ Combine like terms.

So, $5x + 12 = 5x + 8$ has no solution since they have the same coefficient and different constants.

Check

Complete the equation with values that will result in an equation with no solution.

$-3x + 8x - 6 - x = \boxed{} \; x - \boxed{}$

Show your work here

 **Go Online** You can complete an Extra Example online.

Pause and Reflect

How does knowing the structure of equations with infinitely many solutions and equations with no solution help you create these types of equations?

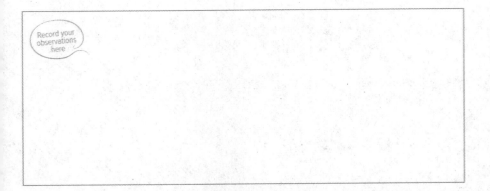
Record your observations here

Talk About It!

How can you verify that the equation
$3(2x + 4) - x = 5x + 8$
has no solution?

Talk About It!

What are some other expressions, other than $5x + 8$, that would result in the equation having no solution?

Pause and Reflect

Create a graphic organizer that will help you study the concepts you learned today in class.

Record your observations here

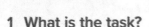

🌐 Apply School

Caden, Lilly, and Amelia are analyzing the expressions $\frac{1}{2}(11x + 24)$ and $-\left(\frac{1}{2}x + 5\right) + 6(x + 8)$, for all values of x. Each student claims the expressions are related according to the results shown in the table. Which student is correct?

Caden	$\frac{1}{2}(11x + 24)$	$=$	$-\left(\frac{1}{2}x + 5\right) + 6(x + 8)$
Lilly	$\frac{1}{2}(11x + 24)$	$>$	$-\left(\frac{1}{2}x + 5\right) + 6(x + 8)$
Amelia	$\frac{1}{2}(11x + 24)$	$<$	$-\left(\frac{1}{2}x + 5\right) + 6(x + 8)$

1 What is the task?

Make sure you understand exactly what question to answer or problem to solve. You may want to read the problem three times. Discuss these questions with a partner.

First Time Describe the context of the problem, in your own words.
Second Time What mathematics do you see in the problem?
Third Time What are you wondering about?

2 How can you approach the task? What strategies can you use?

3 What is your solution?

Use your strategy to solve the problem.

4 How can you show your solution is reasonable?

✏️ **Write About It!** Write an argument that can be used to defend your solution.

📣 **Talk About It!**

For all values of x,

$\frac{1}{2}(11x + 24) <$
$-\left(\frac{1}{2}x + 5\right) + 6(x + 8)$.

Generate two different expressions, A and B, for the right side of the inequality, according to the guidelines below.

$\frac{1}{2}(11x + 24) = A$

$\frac{1}{2}(11x + 24) > B$

Check

Olivia and Harry are analyzing the expressions $\frac{1}{4}(8 + 3x) + 5$ and $-2\left(\frac{3}{4}x + 7\right) + 2\frac{1}{4}x + 24$, for all values of x. Olivia claims that the value of these expressions is always equal. Harry claims that the value of the first expression is always less than the value of the second expression. Which student is correct?

Show your work here

🔵 **Go Online** You can complete an Extra Example online.

Pause and Reflect

What questions do you still have about equations with infinitely many solutions and equations with no solution?

Record your observations here

Practice

Solve each equation. Determine whether the equation has one solution, no solution, or infinitely many solutions. (Examples 1 and 2)

1. $4(x - 8) + 12 = 2(2x - 9)$

2. $3(2k - 5) = 6(k - 4) + 9$

3. $-4y - 3 = \frac{1}{3}(12y - 9) - 8y$

4. $6(3 - 5w) = 5(4 - 2w) - 20w$

Complete each equation so that it has infinitely many solutions. (Example 3)

5. $2x - 7(x + 10) = \boxed{} x - \boxed{}$

6. $12x - x + 8 + 3x = \boxed{} x + \boxed{}$

Complete each equation so that it has no solution. (Example 4)

7. $-15x + 4x + 2 - x = \boxed{} x + \boxed{}$

8. $9(x - 4) - 5x = \boxed{} x - \boxed{}$

Test Practice

9. Multiple Choice Which of the following explains why $\frac{2}{3}(x + 3) = \frac{2}{3}(x - 6)$ has no solution?

Ⓐ The coefficients are different, and the constants are different.

Ⓑ The coefficients are the same, and the constants are the same.

Ⓒ The coefficients are different, and the constants are the same.

Ⓓ The coefficients are the same, and the constants are different.

Apply

10. Three students in the Math Club are analyzing the expressions $\frac{1}{4}(10x + 8)$ and $-(\frac{1}{2}x + 13) + 3(x + 5)$, for all values of x. Each student claims the expressions are related according to the results shown in the table. Which student is correct?

Student 1	$\frac{1}{4}(10x + 8)$	$=$	$-(\frac{1}{2}x + 13) + 3(x + 5)$
Student 2	$\frac{1}{4}(10x + 8)$	$>$	$-(\frac{1}{2}x + 13) + 3(x + 5)$
Student 3	$\frac{1}{4}(10x + 8)$	$<$	$-(\frac{1}{2}x + 13) + 3(x + 5)$

11. Daniel and Fatima are analyzing the expressions below for all values of x. Daniel claims that the value of these expressions is always equal. Fatima claims that the value of the expression on the left is always greater than the value of the expression on the right. Which student is correct?

$$0.6(-7x + 9) \text{ and } 4(x - 3) - (-8 + 8.2x)$$

12. (MP) **Find the Error** A student solved the equation and determined that the solution was -2. Find her error and correct it.

$$1.5x - 2 = -2 + 1.5x$$
$$-2 = -2$$

13. (MP) **Make an Argument** Suppose the solution to an equation is $x = 0$. Explain why it is incorrect to conclude that the equation has no solution.

14. Determine if the statement is *true* or *false*. Justify your response.

An equation will always have at least one solution.

15. (MP) **Identify Structure** What values of a, b, c, and d will make the equation have one solution? Then alter your equation so that it has no solution. Finally, alter the equation again so that it has infinitely many solutions.

$$ax + b = cx + d$$

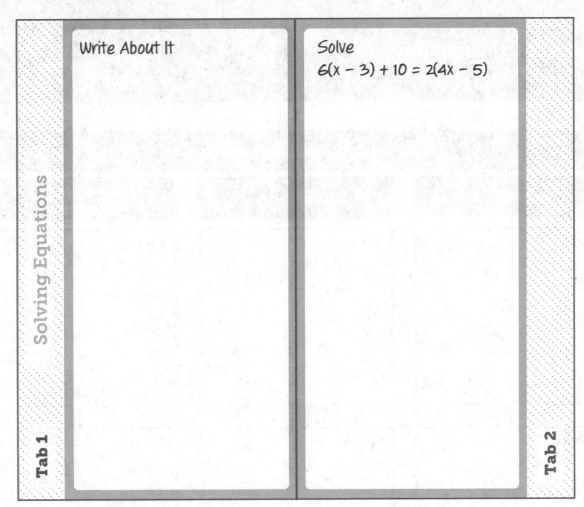

Foldables Use your Foldable to help review the module.

Solving Equations

Tab 1

Write About It

Tab 2

Solve
$$6(x - 3) + 10 = 2(4x - 5)$$

Rate Yourself!   ⭐

Complete the chart at the beginning of the module by placing a checkmark in each row that corresponds with how much you know about each topic after completing this module.

Write about one thing you learned.

_____

_____

_____

_____

Write about a question you still have.

_____

_____

_____

_____

Reflect on the Module

Use what you learned about equations with variables on each side to complete the graphic organizer.

e Essential Question

How can equations with variables on each side be used to represent everyday situations?

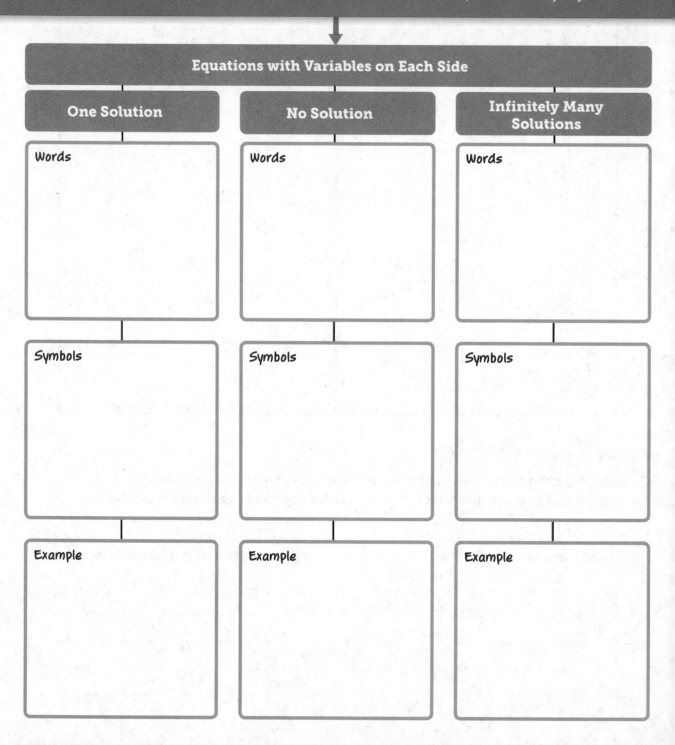

Equations with Variables on Each Side

One Solution	No Solution	Infinitely Many Solutions
Words	Words	Words
Symbols	Symbols	Symbols
Example	Example	Example

Test Practice

1. Equation Editor Solve $3n - 2 = 4n - 6$.

(Lesson 1) $n =$ _____

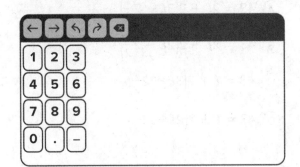

2. Multiple Choice Which of the following represents the solution to the equation $3.4m - 6 = 2.2m - 3.6$? (Lesson 1)

Ⓐ -2

Ⓑ $\frac{1}{2}$

Ⓒ 2

Ⓓ 8

3. Open Response Consider the equation $\frac{2}{3}b + 5 = -\frac{1}{3}b - 4$. (Lesson 1)

A. Solve the equation.

B. Check the solution. Show your work.

4. Multiple Choice A tank has 120 gallons of water and is being drained at a rate of $\frac{1}{2}$ gallon each second. Another tank has 100 gallons of water and is being drained at a rate of $\frac{1}{4}$ gallon each second. Which equation could be used to determine x, the number of seconds, when the two tanks have the same amount of water? (Lesson 2)

Ⓐ $120x + \frac{1}{2} = 100x + \frac{1}{4}$

Ⓑ $120 + \frac{1}{2}x = 100 + \frac{1}{4}x$

Ⓒ $120x - \frac{1}{2} = 100x - \frac{1}{4}$

Ⓓ $120 - \frac{1}{2}x = 100 - \frac{1}{4}x$

5. Open Response Monique wants to rent online movies. Movies Plus charges a one-time fee of $20 plus $5 per movie. Movies-To-Go charges $7 per movie. For how many movies is the cost of renting movies from Movies Plus and Movies-To-Go the same? (Lesson 2)

A. Write an equation to represent the problem, where m is the number of movies rented.

B. Solve the equation.

6. Open Response Cell Phone Plan A charges $25 per month plus $10 per gigabyte of data. Cell Phone Plan B charges $50 per month plus $5 per gigabyte of data. What number of gigabytes of data results in the same cost for one month? (Lesson 2)

7. Equation Editor Solve

$2(6x + 4) - 3x = 5x - 4$. (Lesson 3)

$x =$ _____

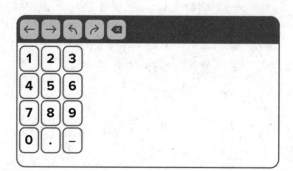

8. Open Response Consider the equation

$-2(3x - 5) - 2x = 2(6 - 3x)$. (Lesson 3)

A. Solve the equation. Show your work.

B. Check the solution. Show your work.

9. Open Response Solve the equation.

$0.4(12 - 2x) = 22.2 - (6x + 7)$. (Lesson 3)

10. Multiple Choice The width of a rectangular table top is $2\frac{1}{2}$ feet shorter than three times its length. If the perimeter of the table top is 43 feet, what is the length, ℓ, of the table top? (Lesson 4)

Which equation represents this situation?

Ⓐ $43 = 2\ell + 2\left(3\ell - 2\frac{1}{2}\right)$

Ⓑ $43 = 2\ell + 2\left(2\ell - 2\frac{1}{2}\right)$

Ⓒ $43 = \ell + 2\left(3\ell - 2\frac{1}{2}\right)$

Ⓓ $43 = \ell + 2\left(2\ell - 2\frac{1}{2}\right)$

11. Table Item Determine whether each equation has one solution, no solution, or infinitely many solutions. (Lesson 5)

	One Solution	No Solution	Infinitely Many Solutions
$4(x + 8) = 2(4 + 2x)$			
$3(2x + 1) = 3 + 6x$			
$2(x + 5) = 5x + 1$			

12. Equation Editor Complete the equation with values that will result in an equation with infinitely many solutions. (Lesson 5)

$7x - 3(x - 1) = \_\_x + \_\_$

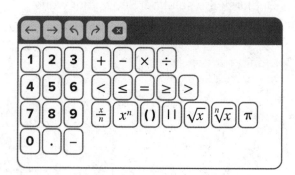

Linear Relationships and Slope

ⓔ Essential Question

How are linear relationships related to proportional relationships?

What Will You Learn?

Place a checkmark (✓) in each row that corresponds with how much you already know about each topic **before** starting this module.

KEY	Before			After		
⬛ — I don't know.　◆ — I've heard of it.　★ — I know it!	⬛	◆	★	⬛	◆	★
finding and interpreting the slope of a proportional relationship						
graphing proportional relationships						
comparing proportional relationships						
finding the slope of a line						
comparing slopes of similar triangles						
writing direct variation equations						
identifying slopes and y-intercepts						
writing equations in slope-intercept form						
graphing linear equations						

📖 Foldables Cut out the Foldable and tape it to the Module Review at the end of the module. You can use the Foldable throughout the module as you learn about linear relationships and slope.

What Vocabulary Will You Learn?

Check the box next to each vocabulary term that you may already know.

☐ constant of proportionality   ☐ linear equation   ☐ slope

☐ constant of variation   ☐ linear relationships   ☐ slope-intercept form

☐ constant rate of change   ☐ rate of change   ☐ slope triangles

☐ corresponding parts   ☐ rise   ☐ solution

☐ direct variation   ☐ run   ☐ unit rate

☐ initial value   ☐ similar figures   ☐ *y*-intercept

Are You Ready?

Study the Quick Review to see if you are ready to start this module.
Then complete the Quick Check.

Quick Review

Example 1
Subtract integers.

Find $-24 - 9$.

$-24 - 9 = -24 + (-9)$ To subtract 9, add -9.

$\qquad = -33$ Simplify.

Example 2
Evaluate expressions.

Evaluate $\dfrac{15 + 5}{9 - 5}$.

$\dfrac{15 + 5}{9 - 5} = \dfrac{20}{4}$ Simplify the numerator and denominator.

$\qquad = 5$ Simplify.

Quick Check

1. A fish was 7 feet below sea level and descended 12 feet. The expression $-7 - 12$ represents this situation. Find $-7 - 12$ to determine the location of the fish compared to sea level.

2. The expression $\dfrac{40 - 13}{22 - 19}$ represents the number of points Cia earned on an assignment. How many points did she earn?

How Did You Do?
Which exercises did you answer correctly in the Quick Check?
Shade those exercise numbers at the right.

Proportional Relationships and Slope

I Can... graph and compare proportional relationships using words, equations, and tables and interpret the unit rate as the slope of the line.

What Vocabulary Will You Learn?
constant rate of change
linear equation
linear relationships
rate of change
slope
solution
unit rate

Explore Rate of Change

Online Activity You will explore how one quantity changes in relation to another quantity.

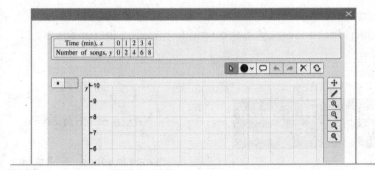

Learn Proportional Relationships

Two quantities are proportional if they vary and have a constant ratio or **unit rate**. The graph of a proportional relationship is a straight line through the origin. Relationships that have straight-line graphs are called **linear relationships.** Proportional relationships can be represented using tables, graphs, words, or equations.

Words	Table			
A linear relationship is proportional when the ratio of y to x is a constant, m.	x	−1	0	1	2
	y	−3	0	3	6

Symbols	Graph
$m = \dfrac{y}{x}$ or $y = mx$, where m is the unit rate and $m \neq 0$.	

Example	
$y = 3x$	$y = 3x$

(continued on next page)

Is the ordered pair (1, 3) a solution to the equation $y = 3x$? Explain why or why not. Name three other solutions.

An equation such as $y = 3x$ is called a linear equation. **A linear equation** is an equation with a graph that is a straight line. Notice that this equation also contains more than one variable. The **solution** of a linear equation consists of two numbers, one for each variable, that makes the equation true.

Learn Unit Rate and Slope

A **rate of change** is a rate that describes how one quantity changes in relation to another quantity. In a linear relationship, the rate of change between any two quantities is the same, or constant. This is called a **constant rate of change.**

The time to download songs is shown in the table. As the number of songs increases by 2, the time in minutes increases by 1.

The unit rate is 2 _____ per _____.

💬 **Talk About It!**

How do the unit rate, slope, and constant rate of change of a proportional linear relationship compare?

Time (minutes), x	Number of Songs, y
0	0
1	2
2	4
3	6

+1 (+2
+1 (+2
+1 (+2

Rate of Change

$$\frac{\text{change in songs}}{\text{change in minutes}} = \frac{2 \text{ songs}}{1 \text{ minute}}$$

In proportional relationships, the unit rate is the slope of the line. **Slope** is the rate of change between any two points on the line.

You can use the points (1, 2) and (2, 4) to find the slope.

$$\frac{\text{change in songs}}{\text{change in minutes}} = \frac{4 - 2}{2 - 1}$$

$$= \frac{\boxed{}}{\boxed{}}$$

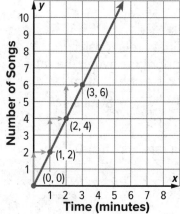

Downloading Songs

⊕ Example 1 Proportional Relationships and Slope

The graph shows the amount of money Ava saved over several weeks.

Find and interpret the slope. Then find the unit rate and compare it to the slope.

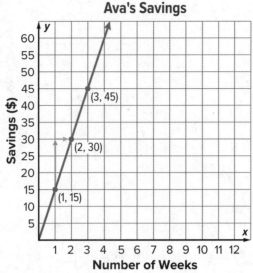

Ava's Savings

Part A Find and interpret the slope.

To find the slope, find the constant rate of change. Choose any two points on the line, such as (1, 15) and (2, 30). Then find the rate of change between the points.

$$\frac{\text{change in savings}}{\text{change in weeks}} = \frac{\$(30 - 15)}{(2 - 1)\ \text{weeks}}$$

$$= \frac{\$\boxed{}}{\boxed{}\ \text{week}}$$

The constant rate of change is $15 per week. So, the slope of the line is $\frac{15}{1}$ or 15. This means that Ava saved $15 every week.

Part B Find the unit rate and compare it to the slope.

Ava saved $_____ every week. So, the unit rate is $15 per week, which is also the slope of the line.

Think About It!

Is this relationship proportional? How do you know?

Talk About It!

How would the slope compare if you had chosen two different points?

Check

The graph shows the relationship between feet and inches.

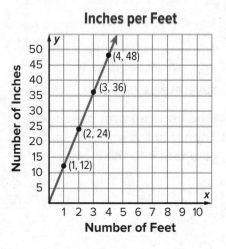

Inches per Feet

Part A

Which of the following is the correct interpretation of the slope?

(A) The slope of the line is $\frac{12}{1}$. This means that there are 12 inches in 1 foot.

(B) The slope of the line is $\frac{1}{12}$. This means that there are 12 inches in 1 foot.

(C) The slope of the line is $\frac{24}{1}$. This means that there are 24 inches in 1 foot.

(D) The slope of the line is $\frac{36}{2}$. This means that there are 36 inches in 2 feet.

Part B

Find the unit rate and compare it to the slope.

Show your work here

🔎 **Go Online** You can complete an Extra Example online.

Pause and Reflect

Reflect on what you have learned about ratios and rates in previous grades. How do the terms *rate of change* and *slope* relate to what you already know?

Record your observations here

🌐 **Example 2** Graph Proportional Relationships

The distance y in miles that a certain cyclist can ride and the time x in hours are in a proportional relationship. This can be represented by the equation $y = 12x$.

Graph the equation. Then find and interpret the slope.

Part A Graph the equation.

Step 1 Make a table of values to find the distance for 0, 1, and 2 hours.

Hours, x	$y = 12x$	Miles, y
0	$y = 12(0)$	
1	$y = 12(1)$	
2	$y = 12(2)$	

Step 2 Graph the ordered pairs.

Graph the ordered pairs (0, 0), (1, 12), and (2, 24) from the table. Then draw a line through the points.

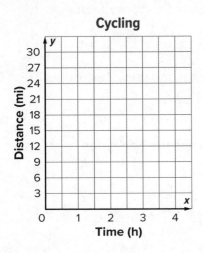

Part B Find and interpret the slope.

In the equation of a proportional linear relationship $y = mx$, m represents the unit rate or slope. The equation $y = 12x$ represents the distance y in miles that the cyclist can ride in x hours.

So, the slope of the line is $\frac{12}{1}$ or 12. This means that the cyclist can ride _____ miles per hour.

> 💭 **Think About It!**
> How does the slope and unit rate compare, for proportional relationships?

> 🗨 **Talk About It!**
> How is the unit rate represented in the table? How is the unit rate represented in the graph?

Check

The number of times y a honeybee can beat its wings and the time in seconds x are in a proportional linear relationship. This situation can be represented by $y = 200x$.

Part A

Graph the equation on the coordinate plane.

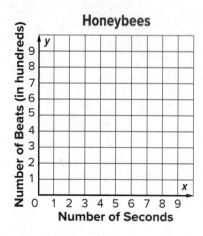

Honeybees

Part B

Which of the following is the correct interpretation of the slope?

(A) The slope of the line is $\frac{1}{100}$. This means the honeybee beats its wings 100 times per second.

(B) The slope of the line is $\frac{100}{1}$. This means the honeybee beats its wings 100 times per second.

(C) The slope of the line is $\frac{1}{200}$. This means the honeybee beats its wings 200 times per second.

(D) The slope of the line is $\frac{200}{1}$. This means the honeybee beats its wings 200 times per second.

Show your work here

🔵 **Go Online** You can complete an Extra Example online.

🌐 Example 3 Graph Proportional Relationships

An online music store charges $3.75 for purchasing three songs. Assume the cost y in dollars is proportional to the number of songs x.

Graph this relationship on the coordinate plane. Then find and interpret the slope.

Part A Graph the relationship.

Step 1 Find the unit rate.

$$\$3.75 \text{ for 3 songs} = \frac{\$3.75}{3 \text{ songs}} \quad \text{Write the rate as a fraction.}$$

$$= \frac{\$1.25}{1 \text{ song}} \quad \text{Simplify.}$$

So, the unit rate is $_____ per song.

Step 2 Make a table of values to find the cost for 1, 2, 3, and 4 songs.

Number of Songs, x	Cost ($), y
1	1.25
2	
3	3.75
4	

Step 3 Graph the line.

Graph the ordered pairs (1, 1.25), (2, 2.50), (3, 3.75), and (4, 5.00) from the table. Then draw a line through the points. Since you cannot purchase part of a song, use a dashed line instead of a solid line.

Online Music

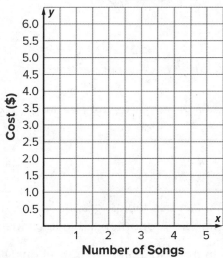

Part B Find and interpret the slope.

What is the unit rate? $_____ per song.

What is the constant rate of change? $_____ per song.

So, the slope of the line is $\frac{1.25}{1}$ or 1.25. This means that the cost is $1.25 per song.

Check

A manatee can swim an average of 10 miles every 2 hours. Assume the distance y in miles is proportional to the number of hours x the manatee swam.

Part A Graph this relationship on the coordinate plane.

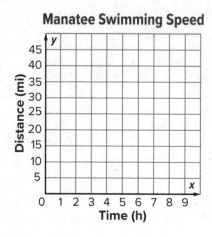

Manatee Swimming Speed

Part B
Find and interpret the slope.

Show your work here

🧭 **Go Online** You can complete an Extra Example online.

Learn Compare Proportional Relationships

You can use tables, graphs, words, or equations to represent and compare proportional relationships.

- In the equation of a proportional linear relationship, $y = mx$, m represents the slope or unit rate.

- In the table, the slope is the constant rate of change or unit rate.

- In the graph, the unit rate or slope is the constant rate of change or the constant ratio $\frac{y}{x}$. You can also find the unit rate from the point $(1, r)$, where r is the unit rate.

(continued on next page)

Words	Equation
The cost y is directly proportional to the number of breakfasts, x.	$y = 2x$
Table	**Graph**

Number of Breakfasts, x	Cost ($), y
0	0
1	2
2	4
3	6
4	8
5	10

+1 ↻ +2
+1 ↻ +2
+1 ↻ +2
+1 ↻ +2
+1 ↻ +2

School Breakfast

Example 4 Compare Proportional Relationships

The distance y in miles that can be covered by a rabbit in x hours can be represented by the equation $y = 35x$. The distance that can be covered by a grizzly bear is shown on the graph.

Which animal is faster? Explain.

Step 1 Find the speed of the rabbit.

The speed is the unit rate. In the equation $y = 35x$, the slope or unit rate is the coefficient of x.
So, the unit rate is _____ miles per hour.

Step 2 Find the speed of the grizzly bear.

In the graph, slope, or unit rate, is the constant rate of change, the constant ratio $\frac{y}{x}$, or the point $(1, r)$.

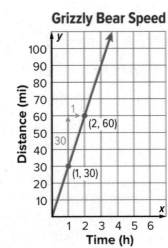

Grizzly Bear Speed

(continued on next page)

Method 1 Find the constant rate of change.

Choose any two points on the line, such as (1, 30) and (2, 60). Then find the rate of change between the points.

$$\frac{\text{change in miles}}{\text{change in hours}} = \frac{(60 - 30) \text{ miles}}{(2 - 1) \text{ hours}} = \frac{\boxed{} \text{ miles}}{\boxed{} \text{ hour}}$$

Method 2 Use the constant ratio.

Write the ratio $\frac{\text{distance}}{\text{time}}$ in simplest form for several points on the line, such as (1, 30) and (2, 60).

$$\frac{\text{distance}}{\text{time}} \rightarrow \frac{30}{1} \qquad \frac{60}{2} = \frac{\boxed{}}{\boxed{}}$$

Method 3 Use the point (1, r).

On the graph, the y-coordinate is 30 when x is _____.

The point (1, 30) represents the unit rate, which is _____ miles per hour.

So, using any of these methods, the grizzly bear travels 30 miles per hour.

Step 3 Compare the unit rates.

The rate for the rabbit is greater than the rate for the grizzly bear.

Since 35 > 30, the _____ is the faster animal.

Check

The equation $y = 1.5x$ represents the relationship between the number of heartbeats y and the time in seconds x for a dog. The graph shows the heartbeats for a cat.

Which animal has a faster heart rate?

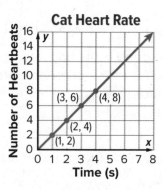

Go Online You can complete an Extra Example online.

Talk About It!

When finding the speed of the grizzly bear, three methods were used. Which method do you prefer? Why?

Example 5 Compare Proportional Relationships

The cost y for computer repairs at Computer Access for x hours is shown in the table. Macro Repair charges $23.50 per hour for computer repairs. Assume that the cost is proportional to the number of hours.

Which company has the lower repair cost? Explain.

Computer Access	
Number of Hours, x	Cost ($), y
2	50
3	75
4	100
5	125

+1 ... +25
+1 ... +25
+1 ... +25

Step 1 Find the unit cost for Computer Access.

Because the relationship is proportional, the constant rate of change is the same as the slope and unit rate.

$$\frac{\textbf{change in cost}}{\text{change in hours}} = \frac{\textbf{\$25}}{\text{1 hour}}$$

As the cost increases by $25, the number of hours increases by 1.

So, the unit cost is $_____ per hour.

Step 2 Find the unit cost for Macro Repair and compare the unit costs.

From the problem, we know that Macro Repair charges $_____ per hour. So, the rate for repairs at Macro Repair is less than the rate for repairs at Computer Access.

Since $23.50 < $25, _____ has the lower repair cost.

Check

Annie's current earnings are shown in the table. She was offered a new job that will pay $7.95 per hour. Assume that her earnings are proportional to the number of hours worked.

Current Earnings	
Hours, x	Money Earned ($), y
2	14.50
3	21.75
4	29.00
5	36.25

Which job pays more each hour?

 Go Online You can complete an Extra Example online.

Pause and Reflect

Write a real-world problem that uses the concepts from today's lesson. Explain how you came up with that problem. Exchange problems with a classmate and solve each other's problem.

Record your observations here

🌐 Apply Utilities

Isabella is comparing rates for two natural gas companies.
For Company A, the cost y for x cubic feet of natural gas is shown in the table. For Company B, the cost y can be represented by the equation $y = 0.53x$, where x represents the number of cubic feet. Which company charges less for 125 cubic feet of natural gas? How much less?

Company A	
Number of Cubic Feet, x	Total Cost ($), y
75	37.50
110	55.00

1 What is the task?

Make sure you understand exactly what question to answer or problem to solve. You may want to read the problem three times. Discuss these questions with a partner.

First Time Describe the context of the problem, in your own words.
Second Time What mathematics do you see in the problem?
Third Time What are you wondering about?

2 How can you approach the task? What strategies can you use?

Record your observations here

3 What is your solution?

Use your strategy to solve the problem.

Show your work here

4 How can you show your solution is reasonable?

✍️ **Write About It!** Write an argument that can be used to defend your solution.

💬 Talk About It!
How did understanding proportional relationships help you solve the problem?

Check

Noah is comparing the cost of two types of silicone wristbands. For Style A, the cost can be represented by the equation $y = 0.65x$, where y represents the cost in dollars and x represents the number of wristbands. For Style B, the costs are shown in the table. Which wristband style costs less for 100 wristbands? How much less?

Style B	
Number of Wristbands, x	Total Cost ($), y
50	23.50
150	70.50

Show your work here

⊳ **Go Online** You can complete an Extra Example online.

📖 **Foldables** It's time to update your Foldable, located in the Module Review, based on what you learned in this lesson. If you haven't already assembled your Foldable, you can find the instructions on page FL1.

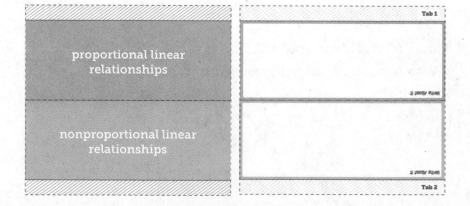

proportional linear relationships

nonproportional linear relationships

Tab 1

Tab 2

Practice

Go Online You can complete your homework online.

1. The graph shows the amount of book sales over several days. Find and interpret the slope. Then find the unit rate and compare it to the slope. (Example 1)

Book Sales

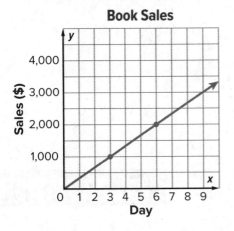

2. The cost y of renting a snowmobile for x hours is a proportional relationship. This can be represented by the equation $y = 33.75x$. Graph the equation. Then find and interpret the slope. (Example 2)

Snowmobile Rental

3. By the end of its fourth week, a movie had grossed $9.2 million. Assume the revenue y in millions of dollars is proportional to the week x. Graph this relationship on the coordinate plane. Then find and interpret the slope. (Example 3)

Movie Sales

4. The amount of power y solar panel A can produce with an area of x square meters can be represented by the equation $y = 1,020x$. The amount of power a solar panel B can produce is shown on the graph. Which solar panel can produce more power? Explain. (Example 4)

Solar Panel B

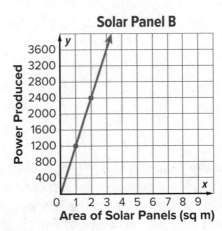

Test Practice

5. Open Response The distance y that Craig biked on x day trips is shown in the table. Rei biked 23.6 miles per day. Assume that the number of miles is proportional to the number of days. Who biked the lower number of miles each day? Explain. (Example 5)

Number of Day Trips	Distance (mi)
2	43.6
3	65.4
4	87.2

Apply

6. Nadia is comparing costs for two brands of garden compost. For Brand A, the cost y for x bags is shown in the table. For Brand B, the cost y can be represented by the equation $y = 1.99x$, where x represents the number of bags. Which brand costs less for 6 bags of compost? How much less?

Number of Bags, x	Total Cost ($), y
3	8.01
5	13.35

7. Explain how the unit rate and slope of a proportional relationship are related to each other.

8. ⓂⓅ **Reason Abstractly** Determine if the statement is *true* or *false*. Justify your response.

The point (1, r) on the graph of a proportional relationship shows the unit rate r.

9. ⓂⓅ **Find the Error** A student graphs the relationship $y = x$ on a coordinate plane. She says the slope is 0 because there is no coefficient. Find her mistake and correct it.

10. ⓂⓅ **Identify Structure** The relationship $y = ax$ is graphed on a coordinate plane. If $a > b$, compare the graph of $y = bx$ to the graph of $y = ax$.

Slope of a Line

I Can... identify the slope of a line and interpret it as the rate of change within the context of the problem.

What Vocabulary Will You Learn?

rise

run

Explore Develop Concepts of Slope

🧭 **Online Activity** You will use Web Sketchpad to explore how to send a point to a target using vertical and horizontal steps.

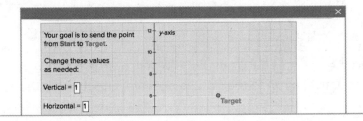

Learn Slope of a Line

The term slope is used to describe the steepness of a line. Slope is the rate of change between any two points on a line. The vertical change (change in *y*-value) is called the **rise** while the horizontal change (change in *x*-value) is called the **run**. So, slope is the ratio of the rise to the run. In linear relationships, the slope is always constant.

$$\text{slope} = \frac{\text{rise} \leftarrow \text{vertical change between any two points}}{\text{run} \leftarrow \text{horizontal change between the same two points}}$$

Slope can be positive or negative. The slope of a line that points upward from, from left to right, is positive, and the slope of a line that points downward, from left to right, is negative.

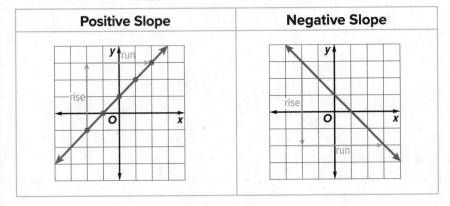

Positive Slope	Negative Slope

Learn Find Slope from a Graph

The slope of a line can be found from a graph by finding the ratio of the rise to the run between any two points on the line.

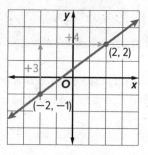

$$\text{slope} = \frac{\text{rise}}{\text{run}} = \frac{\boxed{}}{\boxed{}}$$

← vertical change between (−2, −1) and (2, 2)

← horizontal change between (−2, −1) and (2, 2)

When reading the rise and run from a graph, a rise up is positive, a rise down is negative, a run to the right is positive, and a run to the left is negative.

🧁 **Think About It!**

To travel from the point (2, 4) to the point (3, 6), what is the vertical rise?

💬 **Talk About It!**

How can you tell from the graph that the slope is positive?

🌐 Example 1 Find Slope from a Graph

The graph shows the cost of muffins at a bake sale.

Find the slope of the line.

To calculate the slope, find the ratio of the vertical change (rise) to the horizontal change (run) between any two points on the line. In this case, the points (2, 4) and (3, 6) are used.

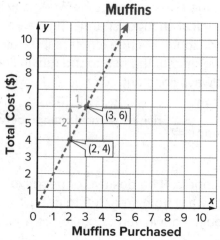

Muffins

$$\text{slope} = \frac{\text{rise}}{\text{run}} \qquad \text{Definition of slope}$$

$$= \frac{\boxed{}}{\boxed{}} \qquad \text{rise} = 2, \text{run} = 1$$

So, the slope of the line is $\frac{2}{1}$ or 2.

Check

Find the slope of the line.

Show your work here

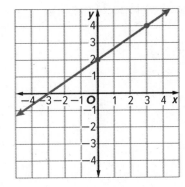

🌐 **Go Online** You can complete an Extra Example online.

🌐 **Example 2** Find Slope from a Graph

The graph shows the amount of water in a leaking bucket over time.

Find the slope of the line.

To calculate the slope, find the ratio of the vertical change to the horizontal change between any two points on the line.
In this case, the points (3, 5) and (9, 1) are used.

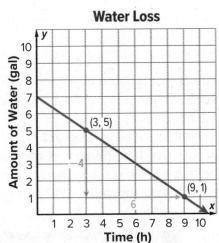

Water Loss

$$slope = \frac{rise}{run}$$ Definition of slope

$$= \frac{\boxed{}}{\boxed{}}$$ rise $= -4$, run $= 6$

$$= \frac{\boxed{}}{\boxed{}}$$ Simplify the ratio.

So, the slope of the line is $\frac{-2}{3}$ or $-\frac{2}{3}$.

Check

Find the slope of the line.

Show your work here

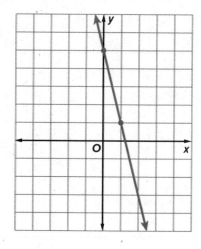

Go Online You can complete an Extra Example online.

Learn Find Slope from a Table

You can determine the slope from a table by finding the ratio of the vertical change, *y*, to the horizontal change, *x*.

The table shows a linear relationship between the balance in a bank account and the number of transactions. The relationship is linear because there is a constant rate of change, or slope. The slope of the relationship shown is −$10 per transaction.

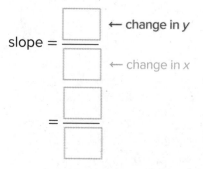

Number of Transactions, *x*	Balance ($), *y*
3	170
6	140
9	110
12	80

+3 ... −30
+3 ... −30
+3 ... −30

$$\text{slope} = \frac{\boxed{}}{\boxed{}} \quad \leftarrow \text{change in } y$$
$$\leftarrow \text{change in } x$$

$$= \frac{\boxed{}}{\boxed{}}$$

Example 3 Find Slope from a Table

The points given in the table lie on a line.

Find the slope of the line. Check your solution.

x	−6	−2	2	6
y	−2	−1	0	1

Choose any two points from the table to find the changes in the x- and y-values. In this example, the points (−6, −2) and (−2, −1) are used.

$$\text{slope} = \frac{\text{change in } y}{\text{change in } x}$$ Definition of slope

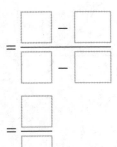   Use the points (−6, −2) and (−2, −1).

So, the slope of the line is $\frac{1}{4}$.

To check, choose two different points from the table and find the slope.

$$\text{slope} = \frac{0 - 1}{2 - 6}$$ Use the points (6, 1) and (2, 0).

$$= \frac{-1}{-4} \text{ or } \frac{1}{4}$$ Simplify.

Check

The points given in the table lie on a line. Find the slope of the line.

x	y
1	3
−7	−1
−15	−5
−23	−9

Go Online You can complete an Extra Example online.

Think About It!

How do you know that this table shows a linear relationship?

Learn Find Slope Using the Slope Formula

You can find the slope of a line from any two points on the line using the slope formula. It does not matter which points you define as (x_1, y_1) and (x_2, y_2). However, the coordinates of both points must be used in the same order.

Words	Model
The slope m of a line passing through points (x_1, y_1) and (x_2, y_2) is the ratio of the difference in the y-coordinates to the difference in the x-coordinates.	
Symbols	
$m = \dfrac{y_2 - y_1}{x_2 - x_1}$, where $x_2 \neq x_1$	

Go Online Watch the animation to learn how to find the slope of a line using the slope formula.

The animation shows how to find the slope of the line using the points (1, 3) and (5, 6).

$m = \dfrac{\text{rise}}{\text{run}}$

$= \dfrac{\text{change in } y}{\text{change in } x}$

$= \dfrac{y_2 - y_1}{x_2 - x_1}$ Slope formula

$= \dfrac{6 - 3}{5 - 1}$ $(x_1, y_1) = (1, 3); (x_2, y_2) = (5, 6)$

$= \dfrac{\boxed{}}{\boxed{}}$ Simplify.

So, the slope of the line is $\dfrac{3}{4}$.

Example 4 Find Slope Using the Slope Formula

Find the slope of the line that passes through $R(1, 2)$, $S(-4, 3)$. Check your solution.

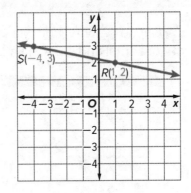

$m = \dfrac{y_2 - y_1}{x_2 - x_1}$ Slope formula

$= \dfrac{\boxed{} - \boxed{}}{\boxed{} - \boxed{}}$ $(x_1, y_1) = (1, 2);$
$(x_2, y_2) = (-4, 3)$

$= \dfrac{\boxed{}}{\boxed{}}$ or $-\dfrac{1}{5}$ Simplify.

So, the slope of the line is $-\dfrac{1}{5}$.

To check, let $(x_1, y_1) = (-4, 3)$ and $(x_2, y_2) = (1, 2)$.

$m = \dfrac{y_2 - y_1}{x_2 - x_1}$ Slope formula

$= \dfrac{2 - 3}{1 - (-4)}$ $(x_1, y_1) = (-4, 3); (x_2, y_2) = (1, 2)$

$= \dfrac{\boxed{}}{\boxed{}}$ Simplify.

Check

Find the slope of the line that passes through $A(-3, 2)$, $B(5, -4)$.

Show your work here

Go Online You can complete an Extra Example online.

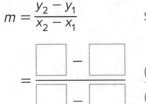**Think About It!**

What is the slope formula?

Talk About It!

A classmate says that the slope of the line is -5. Describe the error they could have made.

Explore Slope of Horizontal and Vertical Lines

Online Activity You will use Web Sketchpad to explore the slopes of horizontal and vertical lines.

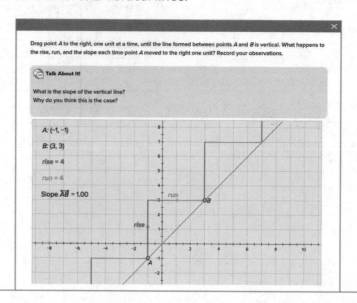

Drag point *A* to the right, one unit at a time, until the line formed between points *A* and *B* is vertical. What happens to the rise, run, and the slope each time point *A* moved to the right one unit? Record your observations.

Talk About It!

What is the slope of the vertical line?
Why do you think this is the case?

A: (–1, –1)
B: (3, 3)
rise = 4
run = 4
Slope $\overline{AB}$ = 1.00

Learn Zero and Undefined Slope

Horizontal lines have a slope of _____.

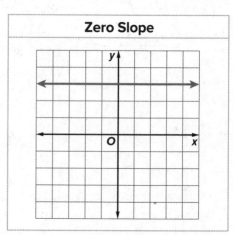

Vertical lines have an _____ slope.

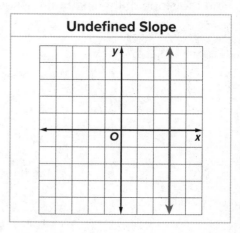

Example 5 Zero Slope

Find the slope of the line that passes through $A(-3, 4)$, $B(2, 4)$.

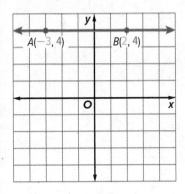

$m = \dfrac{y_2 - y_1}{x_2 - x_1}$ Slope formula

$= \dfrac{\boxed{} - \boxed{}}{\boxed{} - \boxed{}}$ $(x_1, y_1) = (2, 4);$
$(x_2, y_2) = (-3, 4)$

$= \dfrac{\boxed{}}{\boxed{}}$ or 0   Simplify.

So, the slope of the line is 0.

Check

Find the slope of the line that passes through $A(-4, 5)$, $B(2, 5)$.

Show your work here

Go Online You can complete an Extra Example online.

Think About It!
How can you describe the steepness of a horizontal line?

Talk About It!
Why is the slope zero?

Talk About It!
How can you determine that the slope is zero without using the slope formula?

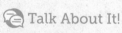

Think About It!

What do you notice about the coordinates of the two points?

Talk About It!

Why is the slope undefined?

Talk About It!

How could you determine that the slope is undefined without using the slope formula?

Example 6 Undefined Slope

Find the slope of the line that passes through $T(1, 3)$, $U(1, 0)$.

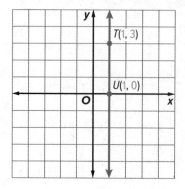

$$m = \frac{y_2 - y_1}{x_2 - x_1} \qquad \text{Slope formula}$$

$$= \frac{0 - 3}{1 - 1} \qquad (x_1, y_1) = (1, 3);\ (x_2, y_2) = (1, 0)$$

$$= \cancel{\frac{3}{0}} \qquad \text{The slope is undefined.}$$

So, the slope of the line is _____.

Check

Which of the following represents the slope of the line that passes through $L(-2, 3)$, $M(-2, 8)$?

(A) 0

(B) 5

(C) $-\frac{5}{4}$

(D) The slope is undefined.

Show your work here

Go Online You can complete an Extra Example online.

🌐 Apply Income

Wesley and Jane both earn money mowing lawns in the summer, and wanted to keep track of their earnings. Wesley tracked his earnings with a graph, while Jane tracked her earnings with a table. Who is earning money at a faster rate? How much more per lawn does that person earn?

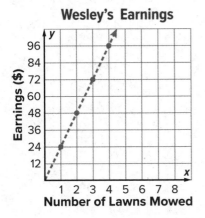

Wesley's Earnings

Earnings ($) vs. Number of Lawns Mowed

Jane's Earnings			
Number of Lawns Mowed, x	4	6	8
Earnings ($), y	64	96	128

1 What is the task?

Make sure you understand exactly what question to answer or problem to solve. You may want to read the problem three times. Discuss these questions with a partner.

First Time Describe the context of the problem, in your own words.
Second Time What mathematics do you see in the problem?
Third Time What are you wondering about?

2 How can you approach the task? What strategies can you use?

3 What is your solution?

Use your strategy to solve the problem.

4 How can you show your solution is reasonable?

✏️ **Write About It!** Write an argument that can be used to defend your solution.

💬 **Talk About It!**

In the graph of Wesley's earnings, why is the rise 24 and not 2?

Check

Lisa and Greg were comparing the gas mileage on their cars. Lisa recorded her gas mileage in a table, while Greg recorded his gas mileage on a graph. Assume the points lie on a line. Whose car gets better gas mileage?

Gas (gal), x	4	8	12	16
Distance (mi), y	108	216	324	432

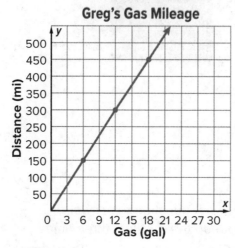

Greg's Gas Mileage

Show your work here

🔎 **Go Online** You can complete an Extra Example online.

Pause and Reflect

Compare and contrast each of the methods used to find slope: from a graph, from a table, using the slope formula.

Record your observations here

Practice

🔁 **Go Online** You can complete your homework online.

1. The graph shows the depth in feet of snow after each two-hour period during a snowstorm. Find the slope of the line. (Example 1)

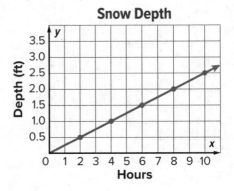

2. The graph shows the amount of money left after buying video games. Find the slope of the line. (Example 2)

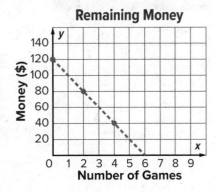

3. The points given in the table lie on a line. Find the slope of the line. (Example 3)

x	−1	2	5	8
y	3	−1	−5	−9

Find the slope of the line that passes through each pair of points. (Examples 4–6)

4. $M(3, 5)$, $N(2, 6)$

5. $G(−3, 2)$, $H(7, 2)$

6. $E(6, 8)$, $F(6, −2)$

Test Practice

7. **Multiple Choice** The points given in each table lie on lines. Which table, when graphed, would show a negative slope?

Ⓐ

x	−2	3	8	13
y	−2	−1	0	1

Ⓒ

x	3	5	6	8
y	8	0	−4	−12

Ⓑ

x	−1	1	3	5
y	−2	0	2	4

Ⓓ

x	−13	−9	−5	−1
y	−4	−2	0	2

Apply

8. Malik and Mila are both mountain climbing. The graph shows Malik's altitude at various points during the climb. The table shows Mila's altitude. Who is climbing at a faster rate? How much faster does that person climb per hour?

Mila's Ascension	
Time (h), x	Altitude (ft), y
0	0
1	721
2	1,442

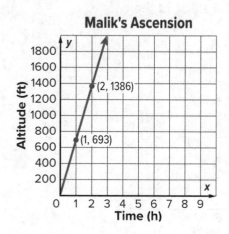

Malik's Ascension

9. Give three points that lie on a line with a slope of $-\dfrac{2}{5}$.

10. Explain why, when using the slope formula, it does not matter which point is (x_1, y_1) or (x_2, y_2).

11. **MP** **Find the Error** A student finds the slope of the line that passes through the points $(-3, 8)$ and $(2, -4)$. Find the mistake and correct it.

$$m = \frac{-3 - 8}{2 - (-4)}$$
$$m = \frac{-11}{6}$$

12. **MP** **Identify Structure** Without computing, how can you tell by looking at the ordered pairs that a line will be horizontal or vertical?

Similar Triangles and Slope

I Can... identify similar triangles that fall on the same line in a coordinate plane and show that the slopes of the lines are equal.

What Vocabulary Will You Learn

corresponding parts

similar figures

slope triangles

Explore Right Triangles and Slope

Online Activity You will use Web Sketchpad to explore why the slope of a line is the same between any two points on a non-vertical line.

> Press *Show Line*. Drag point *B* to different locations. What do you notice about point *B* as it moves? Record your observations.
>
> **Talk About It!**
>
> Press *Show Slope* to see the slope of line *AB*. How does the slope compare to the slope you found?
> As you drag point *B*, what happens to the slope? How does this compare to your earlier prediction?
>
> CB = 3.00
> AC = 6.00
>
> Show Line
> Show Slope

Learn Similar Triangles

When figures have the same shape but not necessarily the same size, they are called **similar figures**. Similar figures have **corresponding parts**.

Words	Model
If two triangles are similar, then their corresponding angles are congruent and the ratios of their corresponding sides are proportional.	

Symbols

$\angle A \cong \angle X$, $\angle B \cong \angle Y$, $\angle C \cong \angle Z$

$$\frac{AB}{XY} = \frac{BC}{YZ} = \frac{AC}{XZ}$$

Talk About It!

The two triangles shown are similar. How do the side lengths of the small triangle compare to the side lengths of the larger triangle?

💬 **Talk About It!**

How can you use the slope triangles to find the slope of the line?

Learn Similar Triangles and Slope

Triangle *ABC* and △*BDE* are both right triangles and they fall on the same line on the coordinate plane. These right triangles are called **slope triangles**. Slope triangles are similar, so their corresponding sides are proportional.

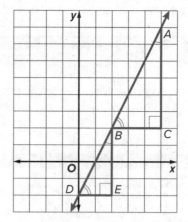

💬 **Talk About It!**

Is the slope of the line the same no matter which slope triangles are used? Explain. Draw other slope triangles to support your explanation.

The vertical and horizontal sides of the slope triangles are the same as the rise and the run of the line. You can use the properties of similar triangles to show the ratios of the rise to the run for each triangle are equal.

$$\frac{AC}{BE} = \frac{BC}{DE}$$ 　Corresponding sides are proportional.

$$\frac{BE}{BC} \cdot \frac{AC}{BE} = \frac{BC}{DE} \cdot \frac{BE}{BC}$$ 　Multiplication Property of Equality

$$\frac{AC}{BC} = \frac{BE}{DE}$$ 　Simplify.

$$\frac{6}{3} = \frac{4}{2}$$ 　Subsitute the rise and run for each slope triangle.

$$\frac{\boxed{}}{\boxed{}} = \frac{2}{1}$$ 　Simplify.

Since the ratios $\frac{AC}{BC}$ and $\frac{BE}{DE}$ are equal, the slope is the same anywhere on the line.

(continued on next page)

Words	Graph

The ratio of the rise to the run of two slope triangles formed by a line is equal to the slope of the line.

Example	

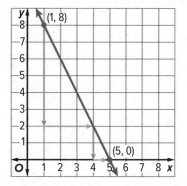

Larger Triangle:

$$\frac{\text{rise}}{\text{run}} = \frac{-6}{3}, \text{ or } \boxed{}$$

Smaller Triangle:

$$\frac{\text{rise}}{\text{run}} = \frac{-2}{1}, \text{ or } \boxed{}$$

Slope of the Line:

$$m = \frac{-8}{4}, \text{ or } \boxed{}$$

 Go Online Watch the video to learn why the slope is the same between any two points on a non-vertical line.

The video shows how to use slope triangles *ABC* and *CDE* to find the slope of the line.

$$\frac{AB}{BC} = \frac{\boxed{}}{\boxed{}}$$

$$\frac{CD}{DE} = \frac{8}{6}$$

$$= \frac{\boxed{}}{\boxed{}}$$

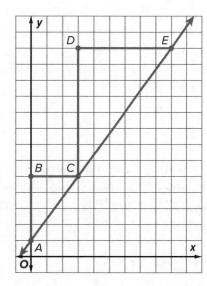

Example 1 Compare Slopes of Similar Triangles

The graph of line _t_ is shown. Use the similar slope triangles to compare the slope of segment _AC_ and the slope of the segment _CE_.

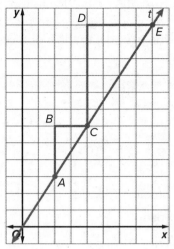

Step 1 Find the slope of segment _AC_.
Use triangle _ABC_ to find the ratio of the rise to the run.

$$\frac{AB}{BC} = \frac{\boxed{}}{\boxed{}} \qquad \text{rise} = 3, \text{run} = 2$$

So, the slope of segment _AC_ is $\frac{3}{2}$.

Step 2 Find the slope of segment _CE_.
Use triangle _CDE_ to find the ratio of the rise to the run.

$$\frac{CD}{DE} = \frac{\boxed{}}{\boxed{}} \begin{array}{l} \leftarrow \text{rise} \\ \\ \leftarrow \text{run} \end{array}$$

$$= \frac{\boxed{}}{\boxed{}} \qquad \text{Simplify.}$$

The slope of segment _CE_ is $\frac{3}{2}$. Since $\frac{AB}{BC} = \frac{CD}{DE}$, the slopes of each similar triangle are the same.

Check

The graph of line _d_ is shown. Use the similar slope triangles to compare the slope of segment _FJ_ and the slope of segment _JK_.

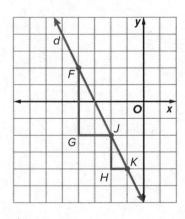

🌐 **Go Online** You can complete an Extra Example online.

 Talk About It!

What is the slope of line _t_? How does it compare to the slope of segment _AC_ and the slope of segment _CE_?

💬 **Talk About It!**

How can you use the properties of similar triangles to explain why the slope is the same between any two distinct points on a line?

Example 2 Verify Slopes Using Slope Triangles

The pitch of a roof refers to the slope of the roof line.

Choose two points on the roof and find the pitch of the roof shown. Then verify that the pitch is the same by choosing a different set of points.

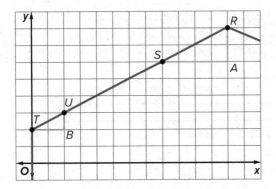

Think About It!
Which two points will you choose to find the pitch of the roof?

Step 1 Use points *T* and *U* to draw slope triangle *TUB*.

slope $= \dfrac{\text{rise}}{\text{run}}$ Definition of slope

$= \dfrac{\boxed{}}{\boxed{}}$ rise = 1, run = 2

So, the pitch of the roof is $\dfrac{1}{2}$.

Step 2 Verify that the pitch is the same using two other points, such as *S* and *R*. Draw slope triangle *SRA*.

slope $= \dfrac{\text{rise}}{\text{run}}$ Definition of slope

$= \dfrac{\boxed{}}{\boxed{}}$ rise = 2, run = 4

$= \dfrac{\boxed{}}{\boxed{}}$ Simplify.

Since the slope is $\dfrac{1}{2}$, the pitch is the same.

Talk About It!
Why is the slope of segment *TU* equal to the slope of segment *SR*?

Check

The plans for a skateboard ramp are shown. Use two points to determine the slope of the ramp. Then verify that the slope is the same by choosing a different set of points.

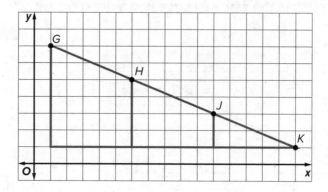

Show your work here

🧭 **Go Online** You can complete an Extra Example online.

Pause and Reflect

How will you study the concepts in today's lesson? Describe some steps you can take.

Record your observations here

Practice

Go Online You can complete your homework online.

1. The graph of line *m* is shown. Use the similar slope triangles to compare the slope of segment *RT* and *TV*. (Example 1)

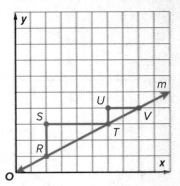

2. The plans for a zipline are shown. Use two points to determine the slope of the zipline. Then verify that the slope is the same by choosing a different set of points. (Example 2)

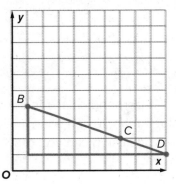

3. Name the slope triangles shown in the graph. What is the slope of the line?

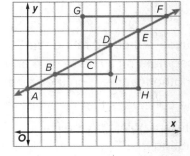

4. Draw two slope triangles on the line. Determine the slope of the line.

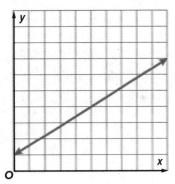

Test Practice

5. **Multiselect** The graph shows similar slope triangles on a line. Select all of the statements that are true.

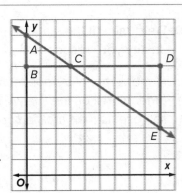

☐ The slope of the line is negative.

☐ The slopes of each triangle are the same because they lie on the same line.

☐ Triangle *CDE* has a greater slope because the triangle is larger.

☐ The slope of each triangle is $\frac{2}{3}$.

☐ The slope of the line is positive.

Apply

6. Lines *r* and *s* are parallel, meaning they will never intersect. Draw similar slope triangles on each line and find the slope of each line. What conclusion can you draw about the slopes of parallel lines?

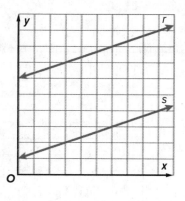

7. Lines *m* and *n* are perpendicular, meaning they form a right angle. Draw similar slope triangles on each line and find the slope of each line. What conclusion can you draw about the slopes of perpendicular lines?

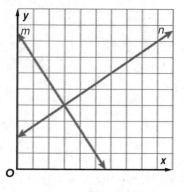

8. ⓂⓅ **Be Precise** How are slope triangles, corresponding sides, ratios, and $\frac{\text{rise}}{\text{run}}$ related?

9. ⓂⓅ **Find the Error** A student found the slope of one segment on a line to be 4 and the slope of another segment on the same line to be $\frac{8}{2}$. He concludes that the slope is different at different points on the line. Correct his thinking.

10. Determine if the statement is *true* or *false*. Justify your response.

If one vertex of a slope triangle is at (0, 0), the other vertex on the line will be the point that represents the simplified unit rate or slope.

11. ⓂⓅ **Identify Structure** Does the placement of the slope triangles matter when finding slope of a line? Explain your reasoning.

Direct Variation

I Can... derive the equation $y = mx$ from the slope formula and use direct variation equations to represent and solve real-world and mathematical problems.

What Vocabulary Will You Learn?

constant of proportionality

constant of variation

direct variation

Explore Derive the Equation $y = mx$

Online Activity You will explore how to use the slope formula to derive the equation $y = mx$.

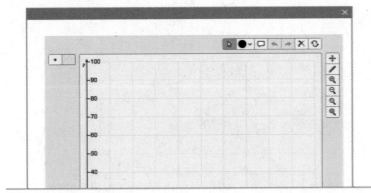

Learn Direct Variation

When the ratio of two variable quantities is constant, a proportional linear relationship exists. This proportional linear relationship is called a **direct variation**. The constant ratio is also called the **constant of variation** or the **constant of proportionality**. In the direct variation equation, $y = mx$, m represents the constant of variation, the constant of proportionality, the slope, and the unit rate.

Words	Example
A direct variation is a linear relationship in which the ratio of y to x is a constant, m. We say y varies directly with x.	$y = 2x$
Symbols	**Graph**
$m = \dfrac{y}{x}$ or $y = mx$, where m is the constant of variation and $m \neq 0$.	

(continued on next page)

The slope of the graph of $y = mx$ is m. Since $(0, 0)$ is one solution of $y = mx$, the graph of a direct variation relationship always passes through the origin.

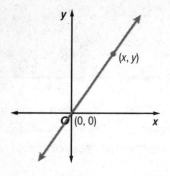

You can use the slope formula to derive the direct variation equation.

$$\frac{y_2 - y_1}{x_2 - x_1} = m \qquad \text{Slope formula}$$

$$\frac{y - \boxed{}}{x - \boxed{}} = m \qquad (x_1, y_1) = (0, 0); (x_2, y_2) = (x, y)$$

$$\frac{\boxed{}}{\boxed{}} = m \qquad \text{Simplify.}$$

$$y = \boxed{} \qquad \text{Multiplication Property of Equality}$$

Pause and Reflect

Did you struggle with any of the concepts in this Learn? How do you feel when you struggle with math concepts? What steps can you take to understand those concepts?

Record your observations here

Example 1 Write Direct Variation Equations From Graphs

The cost y of gymnastics lessons varies directly with the number of sessions x as shown in the graph.

Write a direct variation equation to represent this relationship. Then identify the constant of variation and interpret its meaning.

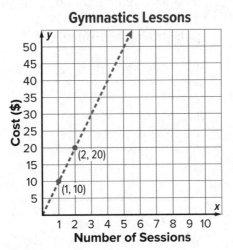

Gymnastics Lessons

Cost ($)

(2, 20)

(1, 10)

Number of Sessions

Think About It!

How do you know that this relationship is a direct variation?

Part A Write a direct variation equation.

Step 1 Find the slope m using the graph.

Choose any two points on the line and find the ratio of the rise to the run between the two points.

$$\text{slope} = \frac{\text{rise}}{\text{run}} \qquad \text{Definition of slope}$$

$$\text{slope} = \frac{\boxed{}}{\boxed{}}$$

So, the slope of the line is $\frac{10}{1}$ or _____.

Step 2 Use the value of m to write the equation.

$$y = mx \qquad \text{Direct variation equation}$$

$$y = \boxed{} x \qquad \text{Replace } m \text{ with 10.}$$

So, the direct variation equation is $y = 10x$.

Talk About It!

In a direct variation relationship, how are the slope, unit rate, and constant of variation related?

Part B Find the constant of variation and interpret its meaning.

The constant of variation is equal to the slope of the graph. So, the constant of variation is $\frac{10}{1}$ or _____.

This means that the unit rate, or cost per session, is $10.

Check

The time y it takes you to hear thunder varies directly with your distance x from the lightning as shown in the graph.

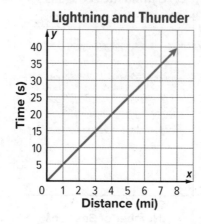

Lightning and Thunder

Part A

Write a direct variation equation to represent this relationship.

Part B

Identify the constant of variation and interpret its meaning.

Show your work here

 **Go Online** You can complete an Extra Example online.

Pause and Reflect

In this Example and Check, you found the constant of variation from a graph. Using what you learned in previous lessons, explain how you could find the constant of variation from words or a table if you are not given the graph.

Record your observations here

Example 2 Write Direct Variation Equations from Words

The cost of bulk peanuts varies directly with the weight of the peanuts. At a local grocery store, 2 pounds of peanuts cost $5.80.

Write a direct variation equation to represent this relationship. Then identify the constant of variation and interpret its meaning.

Part A Write a direct variation equation.

Step 1 Find the unit rate m.

$5.80 for 2 pounds = $\dfrac{\boxed{}}{\boxed{}}$ Write the rate as a fraction.

$= \dfrac{\boxed{}}{1 \text{ pound}}$ Simplify.

So, the unit rate is $2.90 per pound.

Step 2 Use the unit rate to write the equation.

Let y represent the cost ($) of the peanuts and x represent the weight of the peanuts in pounds.

$y = mx$ Direct variation equation

$y = \boxed{}x$ Replace m with $2.90 or 2.9.

So, the direct variation equation is $y = 2.9x$.

Part B Find the constant of variation and interpret its meaning.

The constant of variation is the unit rate.

So, the constant of variation is 2.9.

This means that the cost per pound of peanuts is $_____.

Think About It!

How would you begin solving the problem?

Talk About It!

How can you use the direct variation equation $y = 2.9x$ to determine the cost of 7 pounds of peanuts?

Check

The amount of money Olivia earns varies directly with the number of weeks she works. After 4 weeks, Olivia earned $3,000.

Part A

Write a direct variation equation to represent this relationship.

Part B

Identify the constant of variation and interpret its meaning.

 **Go Online** You can complete an Extra Example online.

Pause and Reflect

Are you ready to move on to the next Example? If yes, what have you learned that you think will help you? If no, what questions do you still have? How can you get those questions answered?

Math History Minute

In 1925, **Elbert Frank Cox (1895–1969)** became the first African American to earn a Ph.D. in mathematics. He was a mathematics professor at West Virginia State College and Howard University. After he retired, Howard University established a scholarship fund in his name to encourage young African Americans to pursue graduate studies in mathematics.

Example 3 Write Direct Variation Equations From Tables

Aubrey is baking oatmeal cookies for the school carnival using the amounts shown in the table. The number of cups of flour varies directly with the number of cups of oats.

Cups of Oats, x	Cups of Flour, y
2	4
4	8
6	12
8	16

Write a direct variation equation to represent this relationship. Then identify the constant of variation and interpret its meaning.

Think About It!

How would you begin solving the problem?

Part A Write a direct variation equation.

Step 1 Find the slope m.

Choose any two points from the table and find the changes in the x- and y- values.

$$\text{slope} = \frac{\text{change in } y}{\text{change in } x} \qquad \text{Definition of slope}$$

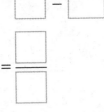

$$\text{slope} = \frac{\square - \square}{\square - \square} \qquad \text{Use the points (2, 4) and (4, 8).}$$

$$= \frac{\square}{\square} \qquad \text{Simplify.}$$

So, the slope of the line is $\frac{2}{1}$ or _____.

Step 2 Use the value of m to write the equation.

$$y = mx \qquad \text{Direct variation equation}$$

$$y = \boxed{} x \qquad \text{Replace } m \text{ with 2.}$$

So, the direct variation equation is $y = 2x$.

Part B Find the constant of variation and interpret its meaning.

The constant of variation is equal to the slope. So, the constant of variation is 2. This means that the unit rate is _____ cups of flour per cup of oats.

Check

The time it takes Madeline to swim laps is shown in the table. The time spent swimming varies directly with the number of laps she swims.

Number of Laps, x	Time (seconds), y
1	45
2	90
3	135
4	180

Part A

Write a direct variation equation to represent this relationship.

Part B

Identify the constant of variation and interpret its meaning.

🔼 **Go Online** You can complete an Extra Example online.

Pause and Reflect

When you first saw this Check, what was your reaction? Did you think you could solve the problem? Did what you already know help you solve the problem?

🌐 Apply Animal Care

A cat's heart can beat 220 times in 2 minutes, nearly twice as fast as a human heart. Assume the number of heartbeats y varies directly with the number of minutes x. Write and solve a direct variation equation to determine how many times a cat's heart beats in 5 minutes.

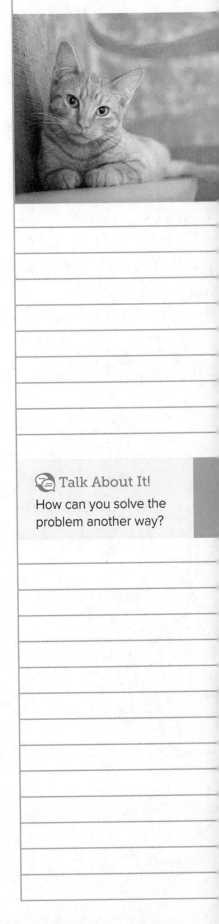

1 What is the task?

Make sure you understand exactly what question to answer or problem to solve. You may want to read the problem three times. Discuss these questions with a partner.

First Time Describe the context of the problem, in your own words.
Second Time What mathematics do you see in the problem?
Third Time What are you wondering about?

2 How can you approach the task? What strategies can you use?

3 What is your solution?

Use your strategy to solve the problem.

💬 Talk About It!
How can you solve the problem another way?

4 How can you show your solution is reasonable?

✍ Write About It! Write an argument that can be used to defend your solution.

Check

A charter bus travels 210 miles in $3\frac{1}{2}$ hours. Assume the distance traveled varies directly with the time traveled. Write and solve a direct variation equation to find how far the bus will travel in 6 hours.

Show your work here

Go Online You can complete an Extra Example online.

Foldables It's time to update your Foldable, located in the Module Review, based on what you learned in this lesson. If you haven't already assembled your Foldable, you can find the instructions on page FL1.

proportional linear relationships

nonproportional linear relationships

Tab 1

Write About It

Tab 2

Write About It

Practice

⟳ Go Online You can complete your homework online.

1. The cost y of movie tickets varies directly with the number of tickets x as shown in the graph. Write a direct variation equation to represent this relationship. Then identify the constant of variation and interpret its meaning. (Example 1)

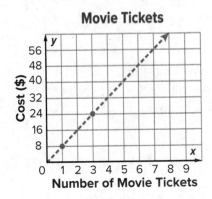

Movie Tickets

2. The number of miles y varies directly with the number of hours x as shown in the graph. Write a direct variation equation to represent this relationship. Then identify the constant of variation and interpret its meaning. (Example 1)

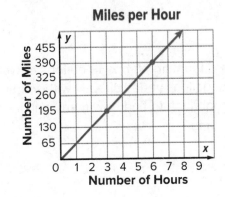

Miles per Hour

3. The cost of paper varies directly with the number of reams bought. Suppose two reams cost $5.10. Write a direct variation equation to represent this relationship. Then identify the constant of variation and interpret its meaning. (Example 2)

4. The amount of flour needed for a recipe varies directly with the number of servings planned. Three servings require $4\frac{1}{2}$ cups of flour. Write a direct variation equation to represent this relationship. Then identify the constant of variation and interpret its meaning. (Example 2)

Test Practice

5. Open Response The distance a bus travels varies directly with time as shown in the table. Write a direct variation equation to represent this relationship. Then identify the constant of variation and interpret its meaning. (Example 3)

Time (h), x	Distance (mi), y
1.5	93.75
3	187.5
4.5	281.25
6	375

Apply

6. Water pressure is measured in pounds per square inch (psi). The number of pounds per square inch y varies directly with the depth x of the water. Write and solve a direct variation equation to determine what the pressure is at a depth of 297 feet.

Depth (ft), x	Pressure (psi), y
66	29
99	43.5
132	58

7. A backyard fountain pumps 18 gallons of water in 4.5 minutes. Assume the number of gallons varies directly with the time. Write and solve a direct variation equation to find how many gallons of water the fountain pumps in 6.5 minutes.

8. Write three ordered pairs that would be found on the line that is graphed by the direct variation equation $y = 3.5x$.

9. The graph of a relationship passes through the points (2, 15.5) and (3, 27). Determine if this is a direct variation relationship. Explain why or why not.

10. MP Find the Error The cost of apps varies directly with the number of apps purchased. Aditi bought four apps for a total of $5.16. She found the direct variation equation below for this relationship. Find her mistake and correct it.

$$y = 5.16x$$

11. How does a constant of variation in a direct variation equation relate to the unit rate?

Slope-Intercept Form

I Can... write equations of the form $y = mx + b$ when given a table, graph, or verbal description.

Explore Derive the Equation $y = mx + b$

Online Activity You will explore how to use the slope formula to derive the equation $y = mx + b$.

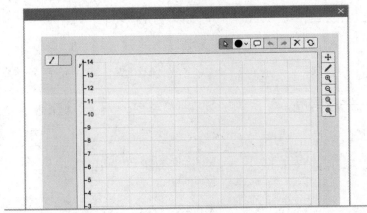

Learn Slope-Intercept Form of a Line

Nonproportional linear relationships can be written in the form $y = mx + b$. This is called the **slope-intercept form**.

slope ⟶

$$y = mx + b$$

⟵ y-intercept

When an equation is written in this form, m is the slope and b is the y-intercept. The **y-intercept** of a line is the y-coordinate of the point where the line crosses the y-axis.

Equation	Graph
$y = \frac{1}{2}x + 1$	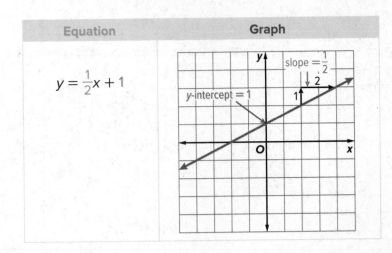

(continued on next page)

In a nonproportional linear relationship, the graph passes through the point $(0, b)$, which is the y-intercept.

You can use the slope formula to derive the equation of a line in slope-intercept form.

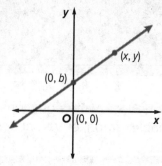

$$\frac{y_2 - y_1}{x_2 - x_1} = m$$ 　　　　Slope formula

$$\frac{y - \boxed{}}{x - \boxed{}} = m$$ 　　　　$(x_1, y_1) = (0, b); (x_2, y_2) = (x, y)$

$$\frac{y - \boxed{}}{\boxed{}} = m$$ 　　　　Simplify.

$$y - b = \boxed{}$$ 　　　　Multiplication Property of Equality

$$y = mx + \boxed{}$$ 　　　　Addition Property of Equality

💭 **Think About It!**

How will using the slope-intercept form of a linear equation help identify the slope and y-intercept?

💬 **Talk About It!**

In the equation $y = \frac{2}{3}x - 4$, why is the y-intercept -4 and not 4?

Example 1 Identify Slopes and y-Intercepts

Identify the slope and y-intercept of the graph of the equation $y = \frac{2}{3}x - 4$.

To identify the slope and y-intercept of the equation, write the equation in the form $y = mx + b$.

$$y = \frac{2}{3}x + \boxed{}$$ 　　　　Write the equation in the form $y = mx + b$.

$$y = mx + b$$ 　　　　$m = \frac{2}{3}, b = -4$

So, the slope of the graph is $\frac{2}{3}$, and the y-intercept is -4.

Check

Identify the slope and y-intercept of the graph of the equation

$y = -\dfrac{2}{5}x - 1$.

Show your work here

Example 2 Write Equations in Slope-Intercept Form

Write the equation of a line in slope-intercept form with a slope of −3 and a y-intercept of −4.

$y = \boxed{}x + b$ Slope-intercept form

$y = \boxed{}x + \left(\boxed{}\right)$ Replace m with −3 and b with −4.

$y = \boxed{}$ Simplify.

So, the equation of the line is $y = -3x - 4$.

Check

Write the equation of a line in slope-intercept form with a slope of 5 and a y-intercept of −7.

Show your work here

 Go Online You can complete an Extra Example online.

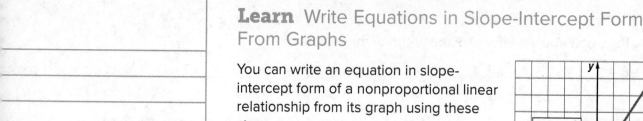

Learn Write Equations in Slope-Intercept Form From Graphs

You can write an equation in slope-intercept form of a nonproportional linear relationship from its graph using these steps.

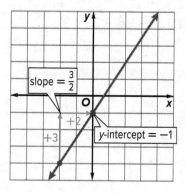

1. Find the location where the line crosses the y-axis to determine the y-intercept.

2. Find the ratio of rise to run to determine the slope.

3. Substitute the values for slope, m, and y-intercept, b, in the equation $y = mx + b$.

$y = $

🍄 **Think About It!**

How would you begin writing the equation?

Example 3 Write Equations in Slope-Intercept Form

Write an equation in slope-intercept form for the graph shown.

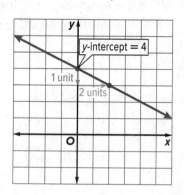

Step 1 Find the y-intercept.

The line crosses the y-axis at (0, 4). So, the y-intercept is _____.

Step 2 Find the slope.

From (0, 4), move down 1 unit and right 2 units to another point on the line.

So, the slope of the line is $\dfrac{\quad}{\quad}$.

Step 3 Write the equation.

Substitute the values for slope, m, and y-intercept, b, in the equation $y = mx + b$.

$y = mx + b$ Slope-intercept form

$y = \boxed{}x + \boxed{}$ Replace m with $-\dfrac{1}{2}$ and b with 4.

So, the equation of the line is $y = -\dfrac{1}{2}x + 4$.

Check

Write an equation in slope-intercept form for the graph shown.

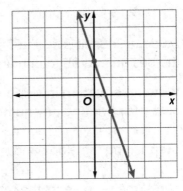

🔵 **Go Online** You can complete an Extra Example online.

Learn Write Equations in Slope-Intercept Form From Verbal Descriptions

When an equation in slope-intercept form applies to a real-world situation, the slope represents the rate of change and the y-intercept represents the **initial value**.

🔵 **Go Online** Watch the animation to learn how to write an equation in slope-intercept form given the following real-world situation.

Bamboo is one of the fastest growing plants on Earth. Suppose a bamboo seedling is 5 centimeters tall and grows at a rate of 6.5 centimeters a day.

Step 1 Find the slope and y-intercept.

$$\text{slope} = \frac{\text{change in height}}{\text{change in time}} = \frac{\boxed{}}{\boxed{}}$$

y-intercept: _____ centimeters

Step 2 Write the equation in slope-intercept form $y = mx + b$.

$$y = \boxed{}\, x + \boxed{}$$

Example 4 Write Equations in Slope-Intercept Form

Student Council is selling T-shirts during spirit week. It costs $20 for the design and $5 to print each shirt.

Write an equation in slope-intercept form to represent the total cost y for printing any number of shirts x.

Step 1 Find the slope and y-intercept.

The slope represents the rate of change or cost per T-shirt. It costs $5 to print each shirt.

The y-intercept represents the initial cost of the design. The one-time charge for the design is $ _____.

So, the slope is $5 and the y-intercept is $20.

Step 2 Write the equation in slope-intercept form $y = mx + b$.

$y = mx + b$ Slope-intercept form

$y = \boxed{}\, x + \boxed{}$ Replace m with the rate of change, 5, and b with the initial cost, 20.

So, the equation that represents the total cost of printing any number of shirts is $y = 5x + 20$.

Check

Faith is saving money in order to purchase a new smartphone. She started out with $30 in her savings account and is able to save an additional $15 a week. Write an equation in slope-intercept form to represent Faith's total savings y for x weeks she puts money toward the purchase of a new smartphone.

Show your work here

Think About It!

What is the slope-intercept form of a line?

Talk About It!

How would the equation $y = 5x + 20$ be altered if the cost to print each shirt was increased to $7?

Go Online You can complete an Extra Example online.

Learn Write Equations in Slope-Intercept Form From Tables

You can write an equation that represents a nonproportional linear relationship in slope-intercept form from a table of values.

Time (s), x	Distance (ft), y
0	10
2	22
4	34
6	46

+2 between x values; +12 between y values

First, determine the slope and y-intercept. Then write the equation in the form $y = mx + b$. The slope, or rate of change, in the table is _____.

The y-intercept, or initial value, is _____. So, the equation is $y = 6x + 10$.

🌐 Example 5 Write Equations in Slope-Intercept Form

Amanda is reading a novel for her Language Arts class. The table shows the number of pages that Amanda has left after a certain number of hours she spent reading.

Write an equation in slope-intercept form that represents the data in the table.

Hours, x	0	2	4	6	8
Pages Left to Read, y	360	280	200	120	40

Step 1 Find the slope.

As the hours increase by 2, the pages left to read decrease by 80.

$$\text{slope} = \frac{\text{change in } y}{\text{change in } x} \qquad \text{Definition of slope}$$

$$= \frac{\boxed{}}{\boxed{}} \qquad \text{Change in } y = -80; \text{ change in } x = 2$$

$$= \frac{\boxed{}}{\boxed{}} \qquad \text{Simplify.}$$

So, the slope is $\frac{-40}{1}$ or -40.

(continued on next page)

💭 Think About It!

How would you begin writing the equation?

💬 Talk About It!

What does the slope represent in the context of this situation?

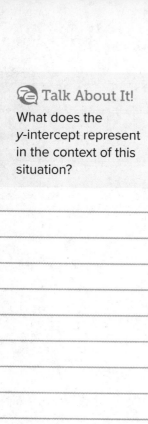

Talk About It!

What does the
y-intercept represent
in the context of this
situation?

Step 2 Find the *y*-intercept.

Find the *y* value when *x* is 0. When *y* is 360, *x* is 0, so the *y*-intercept
(or initial value) is _____.

Step 3 Write the equation.

Substitute the values for slope, *m*, and *y*-intercept, *b*, in the equation.

$y = mx + b$ Slope-intercept form

$y = \boxed{}x + \boxed{}$ Replace *m* with −40 and *b* with 360.

So, the equation that represents the data is $y = -40x + 360$.

Check

The relationship between the data in the table
is linear. Write an equation in slope-intercept
form that represents the data in the table.

Show
your work
here

x	y
−1	6
0	3
1	0
2	−3

Go Online You can complete an Extra Example online.

Pause and Reflect

Did you struggle with writing equations in slope-intercept form when
the information was given in different forms, such as graphs, words,
or tables? If so, what can you do to get help? If not, how could you
explain the process to another student?

Record your
observations
here

🌍 Apply Consumer Science

Amir wants to ship a birthday present to his brother. Express Shipping charges a $5 insurance fee to protect items that are shipped and $0.50 for every ounce the item weighs. Priority Postal's shipping costs are shown in the graph. The present Amir wants to ship weighs 14.2 ounces. Which company charges less to ship the present? How much less?

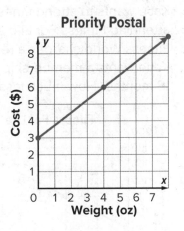

Priority Postal

1 What is the task?

Make sure you understand exactly what question to answer or problem to solve. You may want to read the problem three times. Discuss these questions with a partner.

First Time Describe the context of the problem, in your own words.
Second Time What mathematics do you see in the problem?
Third Time What are you wondering about?

2 How can you approach the task? What strategies can you use?

3 What is your solution?

Use your strategy to solve the problem.

💬 Talk About It!

What does each slope and y-intercept represent in the context of the problem?

4 How can you show your solution is reasonable?

✍️ **Write About It!** Write an argument that can be used to defend your solution.

Check

Katie wants to attend fitness classes at a local gym. The costs of attending Fitness For Life are represented in the graph shown. Fitness World charges a registration fee of $90 plus $8 per month. Katie wants a membership for 18 months. Which gym charges less for 18 months? How much less?

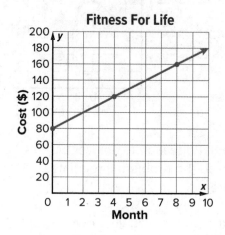

Fitness For Life

Show your work here

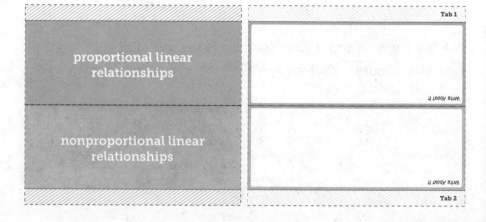

Go Online You can complete an Extra Example online.

Foldables It's time to update your Foldable, located in the Module Review, based on what you learned in this lesson. If you haven't already assembled your Foldable, you can find the instructions on page FL1.

proportional linear relationships

nonproportional linear relationships

Tab 1

Write About It

Write About It

Tab 2

Practice

Go Online You can complete your homework online.

Identify the slope and *y*-intercept of the graph of each equation. (Example 1)

1. $y = \frac{1}{2}x - 5$

2. $y = 3x - 1$

Write the equation of a line in slope-intercept form with each slope and *y*-intercept. (Example 2)

3. slope $= -\frac{1}{3}$, *y*-intercept $= 4$

4. slope $= \frac{3}{2}$, *y*-intercept $= -3$

5. slope $= 4$, *y*-intercept $= -2$

6. slope $= -1$, *y*-intercept $= 6$

7. Write an equation in slope-intercept form for the graph shown. (Example 3)

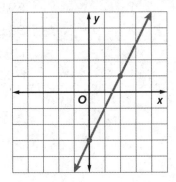

8. The Augello family is driving from Columbus to St. Louis at a constant rate of 65 miles per hour. The distance between the two cities is 420 miles. Write an equation in slope-intercept form to represent the distance *y* in miles remaining after driving *x* hours. (Example 4)

Test Practice

9. The table shows the costs for art show participants, including the $30 registration fee. Write an equation in slope-intercept form that represents the data in the table. (Example 5)

Number of Pieces of Art	Cost ($)
0	30
2	90
4	150
6	210
8	270

10. Multiselect Select all of the statements that are true about the equation $y = -\frac{3}{5}x + 8$.

☐ The slope of the line is negative.

☐ The slope of the line is 8.

☐ The *y*-intercept of the line is 8.

☐ The *y*-intercept of the line is $-\frac{3}{5}$.

☐ The slope of the line is $-\frac{3}{5}$.

Apply

11. While on vacation, Santiago wants to go snorkeling. Coral Snorkeling charges $25 to rent equipment and $30 per hour of boat rental. Sea Water Adventures' costs are shown in the graph. Santiago wants to snorkel for 3 hours. Which company costs less for 3 hours? How much less?

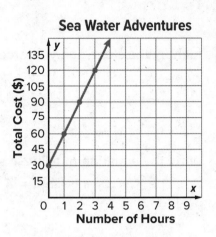

Sea Water Adventures

12. Green's Flowers delivers standard flower arrangements for a $15 delivery fee. Each standard flower arrangement costs $45. The table for Binder's Bouquets shows the costs for different numbers of flower arrangements. Which florist charges less for delivering 8 standard flower arrangements? How much less?

Binder's Bouquets	
Number of Flower Arrangements, x	Total Cost ($), y
0	20.00
1	59.99
2	99.98
3	139.97
4	179.96

13. Write an equation of a line that does not have an x-intercept.

14. ⓂⓅ **Make a Conjecture** Describe what happens to the graph of a line if the slope is doubled.

15. The equation of a line is $y = 3.6x + 2$. What is the rise and run of the slope?

16. ⓂⓅ **Find the Error** Thomas paid a one-time registration fee of $15 and $10 for each cycling class he took. He determined the equation for this relationship is $y = 15x + 10$. Find his error and correct it.

Graph Linear Equations

I Can... interpret the slope and *y*-intercept of a line from an equation of the form $y = mx + b$ in order to graph the line on the coordinate plane.

Learn Graph Equations in Slope-Intercept Form

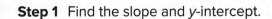 **Go Online** Watch the animation to learn the steps for graphing the equation $y = \frac{2}{3}x + 1$ using the slope and *y*-intercept.

Step 1 Find the slope and *y*-intercept.

Step 2 Graph the *y*-intercept.

Step 3 Use the slope to locate a second point on the line.

Step 4 Draw a line through the points.

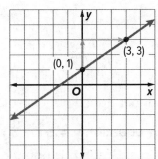

Example 1 Graph Lines Using Slope-Intercept Form

Graph $y = -\frac{2}{3}x + 4$ using the slope and *y*-intercept.

Step 1 Identify the slope and *y*-intercept.

The slope of the line is $-\frac{2}{3}$ and the *y*-intercept is 4.

Step 2 Graph the equation.

Graph the *y*-intercept at (0, 4). Write the slope as $\frac{-2}{3}$. Use it to locate a second point on the line. From the *y*-intercept move down 2 units and right 3 units. Another point on the line is at (3, 2). Draw a line through the points and extend the line.

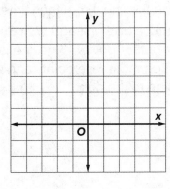

Check

Graph $y = -4x + 5$ using the slope and y-intercept.

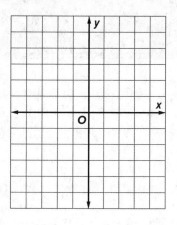

🔂 **Go Online** You can complete an Extra Example online.

🌐 **Example 2** Graph Lines Using Slope-Intercept Form

A typical leopard gecko is 3 inches long at birth and grows at a rate of about $\frac{1}{3}$ inch per week for the first few months.

The equation $y = \frac{1}{3}x + 3$ represents the length y of a gecko after x weeks. Graph this equation.

Step 1 Identify the slope and y-intercept.

$$\text{slope} = \frac{\boxed{}}{\boxed{}} \qquad y\text{-intercept} = \boxed{}$$

So, the slope of the line is $\frac{1}{3}$ and the y-intercept is 3.

Step 2 Graph the equation.

Graph the y-intercept at (0, 3). Use the slope $\frac{1}{3}$ to locate a second point on the line. From the y-intercept move up 1 unit and right 3 units. Another point on the line is at (3, 4). Connect these points and extend the line to the right.

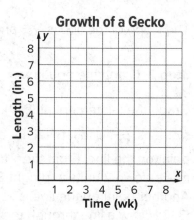

Growth of a Gecko

💭 **Think About It!**

What are the rate of change and initial value?

💬 **Talk About It!**

Use the context of the problem to explain why the line does not extend forever in both directions.

Check

Jayden has written 6 pages of his novel. He plans to write 4 pages per week until he has completed his novel. Graph the equation $y = 4x + 6$ that represents the total number of pages written y in x number of weeks.

Novel Writing

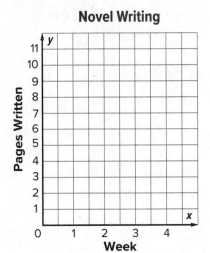

🧭 **Go Online** You can complete an Extra Example online.

Learn Graphs of Horizontal Lines

The graph shows a horizontal line.

All points on a horizontal line have the same y-coordinates.

You can use the slope-intercept form of a line to derive the equation of a horizontal line.

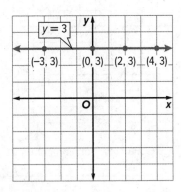

$y = mx + b$ Slope-intercept form

$y = \boxed{}x + b$ The slope m of any horizontal line is 0.

$y = \boxed{}$ Simplify.

So, the equation of a horizontal line can be written as $y = b$, where b is the value of the y-coordinates.

💬 **Talk About It!**
Explain why it makes sense that the equation of the line is $y = 3$.

Example 3 Graph Horizontal Lines

Graph $y = -1$.

The equation $y = -1$ indicates that no matter what the x-values are, the y-values are always -1. Therefore, all points on the line are in the form $(x, -1)$.

Some points along this line are $(-2, -1)$, $(0, -1)$, and $(1, -1)$.

Graph these points. Then draw a line through the points.

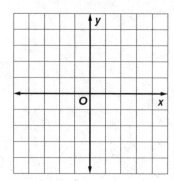

Check

Graph $y = -6$.

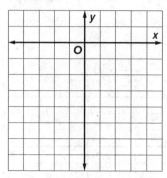

Go Online You can complete an Extra Example online.

Pause and Reflect

Before moving on, recall that the slope of a horizontal line is 0, and the slope of a vertical line is undefined. Work with a partner to develop a memory device that can be used to quickly recall this information. Then share your device with other pairs of students.

Record your observations here

Think About It!

When a line is written in the form $y = b$, what kind of line is it?

Talk About It!

Describe the slope of the line. Does the line have a y-intercept? Explain.

Learn Graphs of Vertical Lines

The graph of a vertical line is shown.

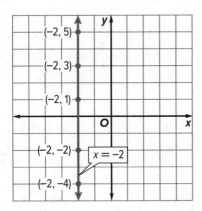

The slope of a vertical line is undefined, so you cannot use the slope-intercept form to derive the equation of a vertical line.

All points on a vertical line have the same *x*-coordinates. In the graph shown, no matter what *y* is, *x* is always −2. So, the equation for the line graphed is $x = -2$.

Therefore, the equation of any vertical line can be written as $x = a$, where *a* is the value of the *x*-coordinates.

Example 4 Graph Vertical Lines

Graph $x = 4$.

The equation $x = 4$ indicates that no matter what the *y*-values are, the *x*-values are always 4. Therefore, all points on the line are in the form (4, *y*).

Some points along this line are (4, −2), (4, 0), and (4, 1).

Graph these points. Then draw a line through the points.

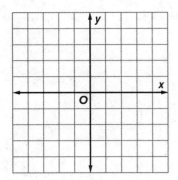

Think About It!

When a line is written in the form $x = a$, what kind of line is it?

Talk About It!

Describe the slope of the line. Does the line have a *y*-intercept? Explain.

Check

Graph $x = 7$.

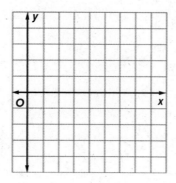

Go Online You can complete an Extra Example online.

Pause and Reflect

Create a graphic organizer that will help you study the concepts you learned today in class.

Record your observations here

🌐 Apply Travel

The Garcia family is driving from Philadelphia to Orlando for vacation. The equation $y = -65x + 1,000$ represents the distance y in miles the Garcia family has left to drive after x hours. Their friends, the Snyders, are meeting them in Orlando, but are driving from Cincinnati. The equation $y = -70x + 900$ represents the distance y in miles the Snyder family has left to drive after x hours. Which family has more miles left to drive after 7 hours? How many more?

🔾 Go Online
Watch the animation.

1 What is the task?

Make sure you understand exactly what question to answer or problem to solve. You may want to read the problem three times. Discuss these questions with a partner.

First Time Describe the context of the problem, in your own words.
Second Time What mathematics do you see in the problem?
Third Time What are you wondering about?

2 How can you approach the task? What strategies can you use?

Record your observations here

3 What is your solution?

Use your strategy to solve the problem.

Show your work here

💬 Talk About It!

What method did you use to solve the problem? Explain why you chose that method.

4 How can you show your solution is reasonable?

🔷 **Write About It!** Write an argument that can be used to defend your solution.

Check

Sefina is tracking the growth of two different plants for an experiment. The growth of Plant A can be represented with the equation $y = \frac{1}{2}x + 1$, where y is the height of the plant after x number of days. The growth of Plant B can be represented with the equation $y = \frac{1}{4}x + 2$. Which plant is taller after 12 days? How much taller?

Show your work here

Go Online You can complete an Extra Example online.

Pause and Reflect

Compare what you learned today with something similar you learned in an earlier module or grade. How are they similar? How are they different?

Record your observations here

Practice

⟲ **Go Online** You can complete your homework online.

Graph each equation using the slope and *y*-intercept. (Example 1)

1. $y = \frac{3}{5}x - 3$

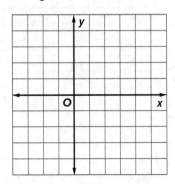

2. $y = -\frac{2}{3}x - 2$

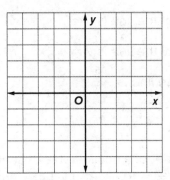

3. The equation $y = \frac{1}{5}x + 3.5$ can be used to find the amount of accumulated snow *y* in inches *x* hours after 5 P.M. on a certain day. Graph this equation. (Example 2)

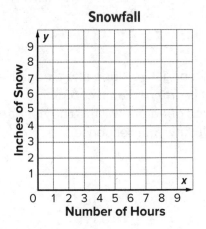

4. Alliyah's gift card balance can be represented by the equation $y = -5x + 50$, where *y* represents the gift card balance after *x* number of days. Graph this equation. (Example 2)

Graph each equation. (Examples 3 and 4)

5. $y = -3$

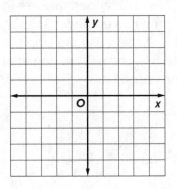

6. $x = 2$

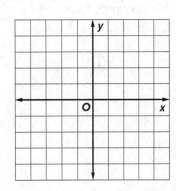

7. Multiselect Select all of the statements that are true for the graph.

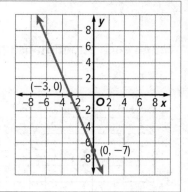

☐ The slope of the line is negative.

☐ The y-intercept is 3.

☐ The equation of the line is $y = 7$.

☐ The equation of the line is $y = -\frac{7}{3}x - 7$.

☐ The slope of the line is $-\frac{3}{7}$.

Apply

8. The altitude y in feet of a hawk that is descending can be represented by the equation $y = -20x + 350$, where x represents the time in minutes. The equation $y = -10x + 400$ represents the altitude y in feet of an eagle after descending x minutes. Which bird is closer to the ground after 8 minutes? How much closer?

9. **Make an Argument** Explain why the equations of vertical lines cannot be in the form $y = mx + b$.

10. A line with a negative slope and a negative y-intercept is graphed on a coordinate plane. Which quadrant will the line *not* pass through? Justify your response.

11. **MP Be Precise** Explain why, when graphing an equation in slope-intercept form, you plot the y-intercept first, rather than the slope.

12. **MP Find the Error** When graphing the equation $y = -3x + 2$, a student plotted the y-intercept at 2, then moved down 3 units and to the left 2 units because the slope is negative, so both rise and run are negative. Find his error and correct it.

📖 **Foldables** Use your Foldable to help review the module.

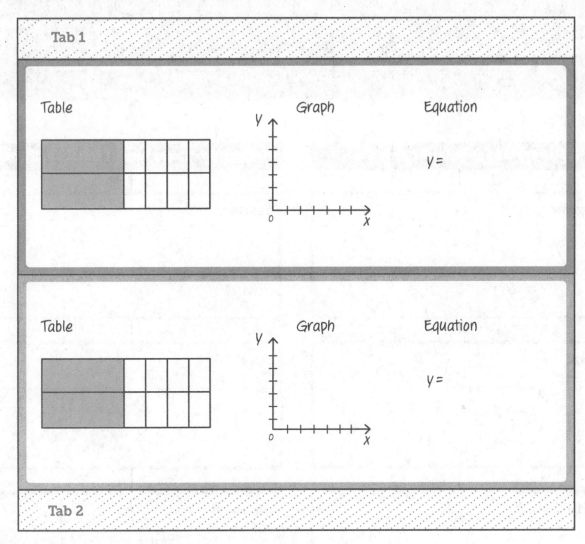

Rate Yourself! ⬛ ◆

Complete the chart at the beginning of the module by placing a checkmark in each row that corresponds with how much you know about each topic after completing this module.

Write about one thing you learned.

Write about a question you still have.

Reflect on the Module

Use what you learned about linear relationships and slope to complete the graphic organizer.

e Essential Question

How are linear relationships related to proportional relationships?

Proportional Relationships

Equation

Slope

y-Intercept

Description of Graph

Nonproportional Linear Relationships

Equation

Slope

y-Intercept

Description of Graph

Test Practice

1. Grid Fluff and Fold charges $2.25 for each load of laundry. **(Lesson 1)**

   A. Draw the graph of the proportional relationship between the two quantities, where x is the number of loads of laundry and y is the total cost.

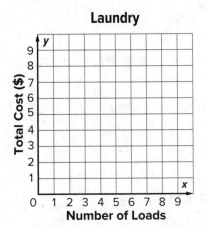

Laundry

   B. Describe how the unit rate is represented in the graph.

2. Open Response Daniella makes apple pies each fall. The cost at the local grocery store for x pounds of apples is shown in the table. What is the least amount of money Daniella will spend for 15 pounds of apples? Assume the relationship is proportional. **(Lesson 1)**

Number of Pounds, x	Total Cost ($), y
2	$4.50
3	$6.75

3. Multiple Choice A turtle is crawling up a hill that rises 6 feet for every horizontal change of 36 feet. Which of the following represents the slope of the hill, as a fraction in simplest form? **(Lesson 2)**

  (A) $\frac{6}{1}$ (C) $\frac{36}{6}$

  (B) $\frac{1}{6}$ (D) $\frac{6}{36}$

4. Open Response The points in the table lie on a line. Compute the slope of the line. **(Lesson 2)**

x	y
6	−3
−2	1
−4	2

5. Multiselect Which statement is true about the graph? Select all that apply. **(Lesson 3)**

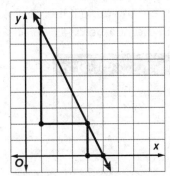

☐ The ratio of the rise to the run of each triangle is the same.

☐ The smaller triangle and the larger triangle shown are similar.

☐ The slope of the line is 2.

☐ The slope of the line is −2.

☐ The corresponding sides of the two triangles are *not* proportional.

6. Open Response The graph of line *m* is shown. Use the similar slope triangles to compare the slope of segment *AD*, the slope of segment *DF*, and the slope of line *m*. (Lesson 3)

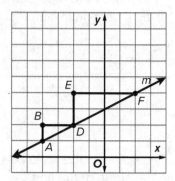

7. Open Response The cost of ground beef varies directly with the number of pounds bought. Suppose 2 pounds cost $8.40. How much would 10.5 pounds of beef cost? (Lesson 4)

8. Table Item Complete the table if the cost *y* varies directly with the number *x*. (Lesson 4)

Number of Snacks, *x*	Cost ($), *y*
2	6
	12
	21
10	

9. Open Response What is the equation of the line that passes through (0, 2) and (−3, 14) in slope intercept form? Justify your reasoning. (Lesson 5)

10. Open Response You are selling tickets. You make $50 for signing up to sell tickets. You also make $0.75 for each ticket you sell. (Lesson 5)

A. Write an equation in slope-intercept form that represents how much money you make, *y*, for selling *x* tickets.

B. How much money will you make for selling 30 tickets?

11. Grid Graph $y = 2x - 4$ using the slope and *y*-intercept. (Lesson 6)

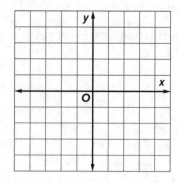

12. Grid Consider the equation $y = -3$. (Lesson 6)

A. Graph $y = -3$.

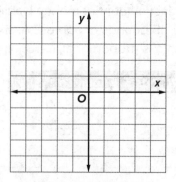

B. Describe the graph of $y = -3$.

Functions

ⓔ Essential Question

What does it mean for a relationship to be a function?

What Will You Learn?

Place a checkmark (✓) in each row that corresponds with how much you already know about each topic **before** starting this module.

KEY			Before			After		
◼ — I don't know. ⬥ — I've heard of it. ★ — I know it!			◼	⬥	★	◼	⬥	★
identifying whether relations are functions								
creating function tables								
graphing linear functions								
constructing linear functions								
comparing two functions								
identifying whether functions are linear or nonlinear								
analyzing qualitative graphs								
sketching qualitative graphs								

📖 Foldables Cut out the Foldable and tape it to the Module Review at the end of the module. You can use the Foldable throughout the module as you learn about functions.

What Vocabulary Will You Learn?

Check the box next to each vocabulary term that you may already know.

☐ function

☐ function table

☐ input

☐ linear function

☐ nonlinear function

☐ output

☐ qualitative graphs

☐ relation

☐ vertical line test

Are You Ready?

Study the Quick Review to see if you are ready to start this module.
Then complete the Quick Check.

Quick Review

Example 1

Name ordered pairs on the coordinate plane.

Name the ordered pair for point B.

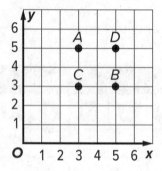

Start at the origin. Move right along the x-axis until you reach 5. Then move up until you reach the y-coordinate of 3. Point B is located at (5, 3).

Example 2

Evaluate expressions given the value of a variable.

Evaluate $7x + 3$ if $x = -2$.

$7x + 3 = 7(-2) + 3$ Replace x with –2.

$\quad\quad = -14 + 3$ Multiply 7 by –2.

$\quad\quad = -11$ Add.

Quick Check

1. Use the graph from Quick Review Example 1. Suppose a park is located at point A. What are the coordinates of the park?

2. Evaluate $3x - 2$ if $x = 4$.

How Did You Do?

Which exercises did you answer correctly in the Quick Check?
Shade those exercise numbers at the right.

Identify Functions

I Can... determine whether or not a relation is a function by identifying the number of outputs assigned to each input.

Explore Relations and Functions

🔎 **Online Activity** You will use Web Sketchpad to explore relations and functions using mapping diagrams.

| Dart! |
| d = 16.0 |
| t = 0.0 |
| 0   5   10   15 |
| Reset |

Learn Relations and Functions

A mathematical relation exists when two mathematical values are related in some way. A **relation** is any set of ordered pairs. The **input** of the relation is the set of x-coordinates. The **output** is the set of y-coordinates. Relations can be represented as shown.

Ordered Pairs
(−2, 3)
(0, −1)
(1, 2)
(3, 1)

Table	
x	y
−2	3
0	−1
1	2
3	1

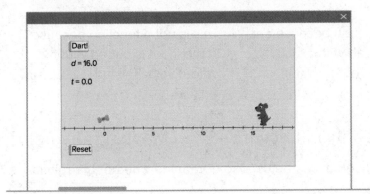

Mapping

Graph

(continued on next page)

A **function** is a specific type of relation that assigns exactly one output to each input.

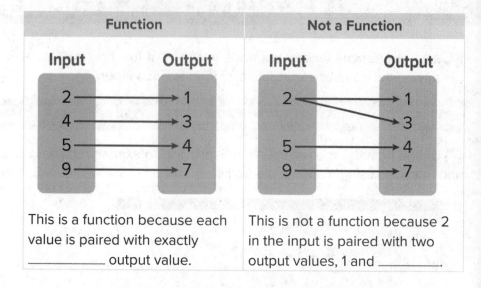

Function		Not a Function	

This is a function because each value is paired with exactly _____ output value.

This is not a function because 2 in the input is paired with two output values, 1 and _____.

Learn Identify Functions Using Mapping Diagrams

A mapping diagram can be used to determine whether the relation is a function.

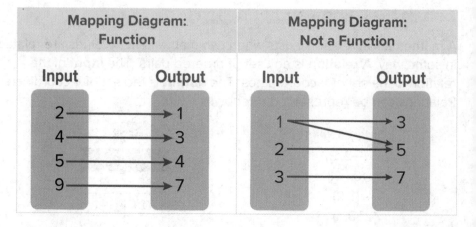

Mapping Diagram: Function		Mapping Diagram: Not a Function	

The mapping diagram labeled "Not a Function" shows the input _____ is mapped to both 3 and 5. Because an input value is mapped to more than _____ output, the relation is not a function.

Example 1 Identify Functions Using Mapping Diagrams

Determine whether the relation is a function. Explain.

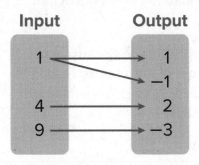

Input **Output**

Check each input value. If an input value is mapped to more than one output value, then the relation is not a function.

So, the relation is not a function because the input value _____ is mapped to more than one output value.

Check

Determine whether the relation is a function. Explain.

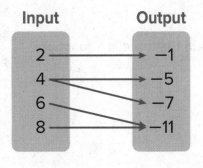

Input **Output**

Show your work here

Go Online You can complete an Extra Example online.

Think About It!

How does a mapping diagram help you determine if a relation is a function?

Talk About It!

How can you alter the relation so that it will be a function?

Learn Identify Functions Using Tables

Tables can be used to represent relations and functions. To determine whether a relation shown in a table is a function, consider how the members of the input correspond to the members of the output.

Function	
Input, x	Output, y
1	2
2	4
3	6
4	8

This is a function because each member of the input corresponds to exactly _____ member of the output.

Not a Function	
Input, x	Output, y
1	2
1	4
3	6
3	8

This is not a function because members of the input correspond to more than one member of the ouput. Specifically, 1 corresponds to both 2 and _____, and 3 corresponds to both 6 and _____.

😯 **Think About It!**

What do you know about the relationship between the input values and the output values in a function?

Example 2 Identify Functions Using Tables

Determine whether the relation shown in the table is a function. Explain.

Input, x	Output, y
−2	4
2	4
−3	9
3	9

Study the table. Are there any input (x) values that correspond to more than one output value? _____

The relation is a function because each input value corresponds to exactly _____ output value.

💬 **Talk About It!**

In this example, some output values correspond to more than one input value. Why is this relation still a function?

Check

Determine whether the relation shown in the table is a function. Explain.

Input, x	Output, y
1	3
2	3
3	3
2	4

Show your work here

Go Online You can complete an Extra Example online.

Learn Identify Functions Using Graphs

The ordered pairs of a relation can be graphed on the coordinate plane. To determine whether the relation shown on a graph is a function, you can apply the **vertical line test**. If, for each value of x, a vertical line passes through no more than one point on the graph, then the graph represents a function. If the line passes through more than one point on the graph, it is *not* a function.

To perform the vertical line test, you can use a pencil or a straightedge to represent a vertical line. Place the pencil at the left of the graph, and then move it to the right across the graph.

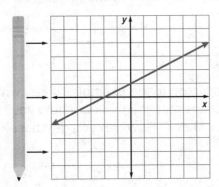

The graph is a function because the vertical line test shows that it always intersects the graph at exactly one point. This means that each input corresponds to _____ one output.

(continued on next page)

Perform the vertical line test on the following graph.

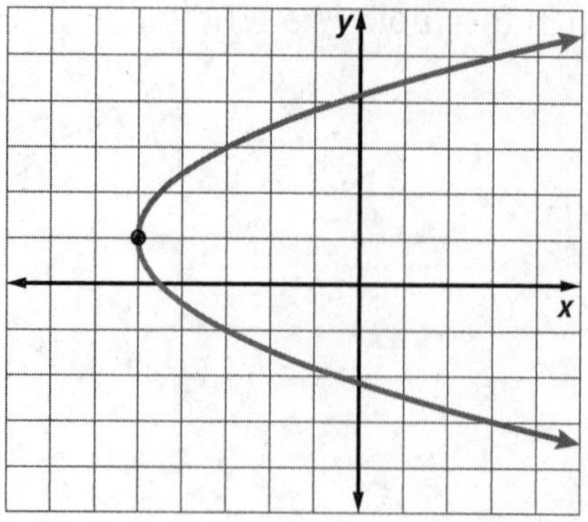

The graph is not a function because the vertical line test shows that it intersects the graph at more than one point. This means that an input corresponds to _____ one output.

For the following graph, refer to the Talk About It!

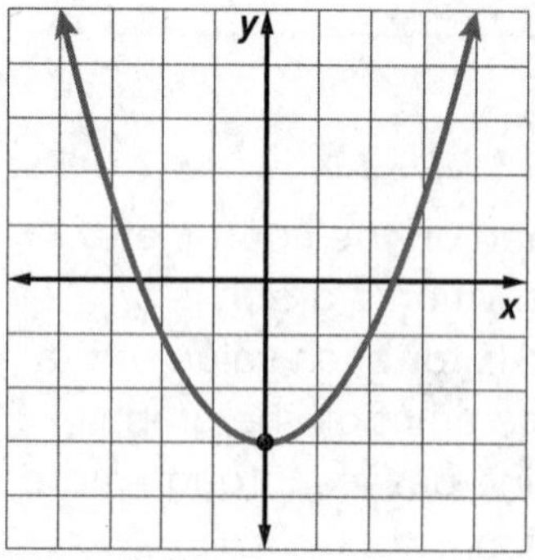

The graph _____ a function because the vertical line test shows that it always intersects the graph at exactly _____ point. This means that each _____ corresponds to exactly one output.

Talk About It!

Before performing the vertical line test, make a conjecture as to whether or not the graph shown is a function. Explain your reasoning. Then perform the vertical line test with a pencil to test your conjecture.

Pause and Reflect

Are you ready to move on to the Example? If yes, what have you learned about functions that you think will help you? If no, what questions do you still have? How can you get those questions answered?

Record your observations here

Example 3 Identify Functions Using Graphs

Determine whether the relation shown in the graph is a function. Explain.

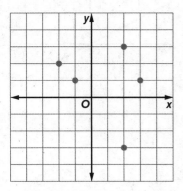

Use the vertical line test. The relation is not a function because the input _____ corresponds to more than one output.

Check

Determine whether the relation shown in the graph is a function. Explain.

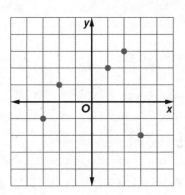

Show your work here

 Go Online You can complete an Extra Example online.

Example 4 Identify Functions Using Graphs

Determine whether the relation shown in the graph is a function. Explain.

Use the vertical line test on the graph of the line. Does the vertical line ever pass through more than one point? _____

The relation is a function because each _____ corresponds to exactly one output.

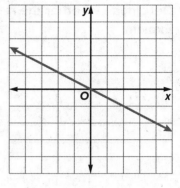

Check

Determine whether the relation shown in the graph is a function. Explain.

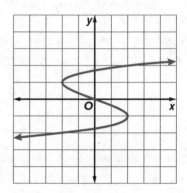

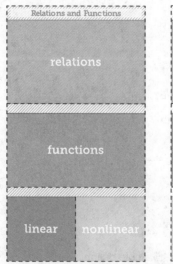

🧭 **Go Online** You can complete an Extra Example online.

___

📒 **Foldables** It's time to update your Foldable, located in the Module Review, based on what you learned in this lesson. If you haven't already assembled your Foldable, you can find the instructions on page FL1.

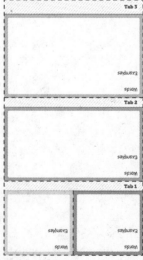

Practice

Determine whether each relation is a function. Explain. (Examples 1-3)

1.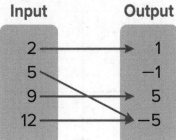

2.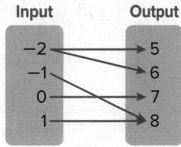

3.

Input, x	Output, y
−10	4
−5	4
0	4
5	4

4.

Input, x	Output, y
1	2
1	3
1	4
1	5

5.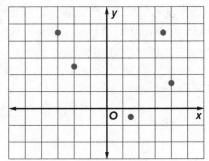

6. **Multiple Choice** Select the statement that correctly explains whether or not the relation shown in the graph is a function. (Example 4)

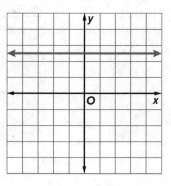

Ⓐ The relation is a function because each input has exactly one output.

Ⓑ The relation is a function because each output has exactly one input.

Ⓒ The relation is not a function because at least one input has more than one output.

Ⓓ The relation is not a function because at least one output has more than one input.

7. **Create** Draw a graph to represent a real-world situation that represents a relation that is a function. Describe the situation and explain why the graph is a function.

8. 🔵 **Make an Argument** Camilla walks at a steady pace for 5 minutes, stops in the same location for 2 minutes, then continues on at a faster pace for 3 more minutes until she reaches her destination. Camilla claims that the relationship between the time spent walking and the distance traveled is not a function because of the time she stood still. Explain why the relationship is a function.

9. Describe a math tool you could use to determine if a relation is a function. Explain how you would use it.

10. 🔵 **Use a Counterexample** If a relation is a function, will reversing the input and output also result in a function? If yes, explain why. If not, give a counterexample.

Function Tables

I Can... generate function tables from function rules and use the sets of ordered pairs to graph the functions.

Explore An Introduction to Function Rules

Online Activity You will use Web Sketchpad to explore how to find outputs from a given input using function rules.

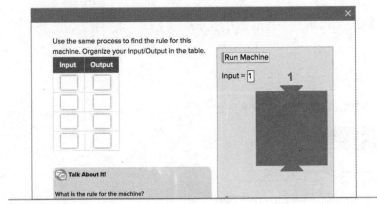

Learn Function Tables

A **function table** is a table that organizes the input and output of a function. You can create a function table from a function rule by substituting appropriate input values into the rule, then simplifying to find the output. An example of a function table is shown.

Function Rule	Function Table		
$y = 2x + 3$	**Input**	**Rule**	**Output**
	x	$2x + 3$	y
	1	$2(1) + 3$	5
	2	$2(2) + 3$	7
	3	$2(3) + 3$	9
	4	$2(4) + 3$	11

Talk About It!
How can you use the function rule $2x + 3$ to determine the output if the input is 7?

Example 1 Create Function Tables

Create a function table for $y = 4x - 1$. Use the input values −2, 0, 2, and 4.

The input values are substituted for x into the function rule. Simplify the expression to find each output.

Input	Rule	Output
x	$4x - 1$	y
−2	$4(-2) - 1$	
0	$4(0) - 1$	
2	$4(2) - 1$	
4	$4(4) - 1$	

So, the values for y when x is −2, 0, 2, and 4 are −9, −1, 7, and 15.

Check

Complete the function table for $y = -4x - 3$.

Input	Output
x	y
−1	
1	
3	
5	

Show your work here

🔾 **Go Online** You can complete an Extra Example online.

Pause and Reflect

How is a function rule represented in a function table?

Record your observations here

🌐 Example 2 Choose Appropriate Input Values

It takes approximately 770 peanuts to produce one jar of peanut butter. The total number of peanuts n is represented by the function $n = 770p$, where p is the number of jars of peanut butter purchased.

Determine appropriate input values for this situation. Then complete a function table for $n = 770p$.

Part A Determine the appropriate input values.

Only _____ numbers make sense for the input because you cannot buy a fraction of a jar or negative jars of peanut butter. Some appropriate input values could be 1, 2, 3, and 4 jars of peanut butter.

Part B Complete a function table.

Substitute the input values 1, 2, 3, and 4 for p into the function $n = 770p$, then simplify to find each output.

Input, p	Output, n
1	
2	
3	
4	

Check

An online photo printing service charges $0.15 per photo and $2.99 for shipping. The total cost c for printing and shipping any number of pictures p can be represented by the function $c = 0.15p + 2.99$.

Part A

What input values for p make sense in this situation?

(A) rational numbers

(B) whole numbers

(C) decimals

(D) irrational numbers

Part B

Complete the function table for $c = 0.15p + 2.99$.

Input, p	Output, c
1	
2	
3	
4	

Show your work here

🌐 **Go Online** You can complete an Extra Example online.

💭 Think About It!

Which variable represents the input value?

💬 Talk About It!

What do the values 4 and 3,080 represent in this situation?

Learn Graph Functions

The graph of a function is the set of ordered pairs, consisting of an input and the corresponding output, that make the equation representing the function true.

A **linear function** is a function in which the graph of the solutions forms a straight line. Therefore, an equation of the form $y = mx$ or $y = mx + b$ is a linear function.

The function $y = x - 1$ is represented as a table, ordered pairs, and a graph below.

Table		Graph

x	y
0	−1
1	0
2	1
3	2

Ordered Pairs

(0, −1), (1, 0), (2, 1), (3, 2)

Pause and Reflect

Compare and contrast each of the different representations (equation, table, graph, ordered pairs) of a linear function.

Record your observations here

Example 3 Graph Linear Functions

Create a function table for the function $y = x + 2$. Then graph the function.

Part A Create a function table.

Select any values for the input x, such as 0, 1, 2, and 3. Substitute these values for x to find the value of y. Then write the corresponding ordered pairs.

x	$x + 2$	y	(x, y)
0	0 + 2		$\left(0, \boxed{}\right)$
1	1 + 2		$\left(1, \boxed{}\right)$
2	2 + 2		$\left(2, \boxed{}\right)$
3	3 + 2		$\left(3, \boxed{}\right)$

Part B Graph the function.

Graph the ordered pairs (0, 2), (1, 3), (2, 4) and (3, 5). Then draw a line that passes through each point.

The line is the complete graph of the function. Each ordered pair of any point on the line is a solution of the equation that represents the function.

(continued on next page)

Think About It!

Which variable represents the input value?

Talk About It!

Is the point (−1, 1) a solution to the function $y = x + 2$? How do you know?

Check

Create a function table for the function $y = -3x + 4$. Then graph the function.

Part A

Create a function table for $y = -3x + 4$.

x	y
−1	
0	
1	
2	

Part B

Graph the function.

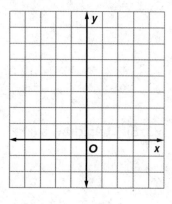

🔵 **Go Online** You can complete an Extra Example online.

Pause and Reflect

Did you ask questions about today's lesson? Why or why not?

Record your observations here

🌏 Apply Transportation

A taxi service charges $2.50 plus $0.75 per mile. The function $y = 0.75x + 2.50$ represents the total cost y, of a taxi ride for any number of miles x. If Adriana rode 10 miles in the taxi, and Alonso rode 15 miles, how much more did Alonso spend on his taxi ride?

Go Online
Watch the animation.

1 What is the task?

Make sure you understand exactly what question to answer or problem to solve. You may want to read the problem three times. Discuss these questions with a partner.

First Time Describe the context of the problem, in your own words.
Second Time What mathematics do you see in the problem?
Third Time What are you wondering about?

2 How can you approach the task? What strategies can you use?

Record your observations here

3 What is your solution?

Use your strategy to solve the problem.

Show your work here

💬 Talk About It!
How could you solve this problem another way?

4 How can you show your solution is reasonable?

🖊 **Write About It!** Write an argument that can be used to defend your solution.

Check

An elephant weighs 200 pounds at birth and gains approximately 2 pounds per day. The function $w = 2d + 200$ represents the weight w of an elephant on a given day d during his first year. How much more does an elephant weigh on Day 60 than it does on Day 5?

Go Online You can complete an Extra Example online.

Foldables It's time to update your Foldable, located in the Module Review, based on what you learned in this lesson. If you haven't already assembled your Foldable, you can find the instructions on page FL1.

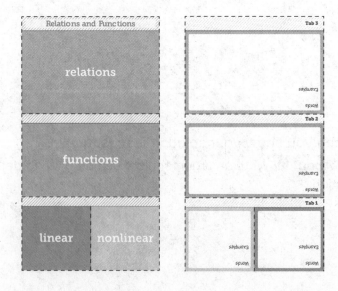

Practice

Go Online You can complete your homework online.

Complete the function table for each function given. (Example 1)

1. $y = 2.5x - 8$

Input, x	Output, y
−5	
0	
5	
10	

2. $y = -5x - 1$

Input, x	Output, y
−2	
−1	
0	
1	

3. $y = \frac{1}{2}x + 3$

Input, x	Output, y
−2	
2	
6	
10	

4. A single-engine plane can travel up to 140 miles per hour. The total number of miles m is represented by the function $m = 140h$, where h is the number of hours traveled. Determine appropriate input values for this situation. Then complete the function table for $m = 140h$. (Example 2)

Input, h	Output, m

Test Practice

5. Create a function table for the function $y = -2x + 1$. Then graph the function. (Example 3)

Input, x	Output, y
−2	
−1	
0	
1	

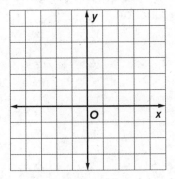

6. **Multiselect** Select all of the possible types of numbers that are appropriate input values for the given situation.

A flower-delivery service charges $39.95 per flower arrangement and $2.99 for delivery. The total cost y is represented by the function $y = 39.95x + 2.99$, where x is the number of flower arrangements.

- ☐ whole numbers
- ☐ integers
- ☐ rational numbers
- ☐ positive integers
- ☐ negative numbers
- ☐ only zero

Apply

7. A baby giraffe is about $6\frac{1}{2}$ feet tall at birth and grows about $\frac{1}{2}$ foot per month for its first year. The function $h = \frac{1}{2}m + 6\frac{1}{2}$ represents the total height h, of a baby giraffe for any number of months m within its first year. How much taller is the giraffe at the end of month 11 than at the end of month 2?

8. A public swimming pool holds 250,000 gallons of water and is being drained at a rate of 200 gallons per minute. The function $g = -200m + 250,000$ represents how many gallons remain in the pool g, after any number of minutes m. How much more water remains after 5 minutes than after 20 minutes?

9. **MP** **Identify Structure** If a linear function is continuous, it has an infinite number of solutions. Which representation(s), equation, table, or graph, show(s) all of the solutions? Explain.

10. **MP** **Justify Conclusions** Lian's truck has a 25 gallon tank and uses 0.05 gallon of gas for every mile driven. When creating a table for the function $y = -0.05x + 25$, Lian argues that he can only use positive rational numbers for the input. Is Lian correct? Justify your answer.

11. Kai completed the function table for the function $y = 3x$ as shown. Find his error and correct it.

Input, x	Output, y
3	1
6	2
9	3

12. **Create** Write a real-world problem in which rational numbers would be appropriate input values for the situation. Then write the equation of the function.

Construct Linear Functions

I Can... write linear functions from graphs, tables, and verbal descriptions by finding the rate of change and initial value.

Learn Rate of Change and Initial Value

A linear function is a function that has an equation in the form $y = mx$ or $y = mx + b$. The graph of a linear function forms a straight line.

The slope, m, of a linear function is also known as the rate of change. The rate of change between any two points in a linear relationship is the same or constant.

The y-intercept, b, of a linear function is also known as the initial value. The initial value is the corresponding y-value when x equals 0.

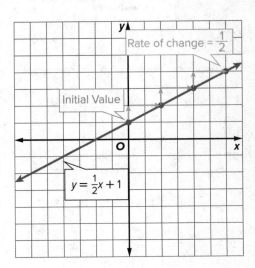

Rate of change = $\frac{1}{2}$

Initial Value

$y = \frac{1}{2}x + 1$

💬 **Talk About It!**

Why do you think the y-intercept is also known as the initial value?

Pause and Reflect

How does this compare to what you have previously learned about proportional and nonproportional relationships?

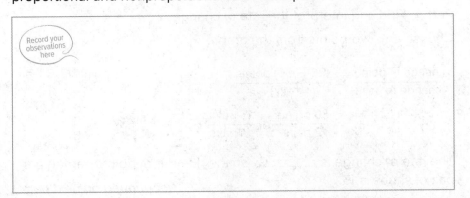

Record your observations here

Learn Construct Linear Functions from Graphs

You can construct a linear function from a graph by determining the rate of change or slope, m, and initial value or y-intercept, b, then writing it in the form $y = mx + b$.

Go Online Watch the animation to see the steps for constructing a linear function from the graph shown.

Step 1 Find the rate of change.

$$\frac{\$300}{4 \text{ months}} = \frac{\$\boxed{}}{\boxed{} \text{ month}}$$

Step 2 Find the initial value.

Initial cost: $\boxed{}$

Step 3 Write the equation of the function.

$$y = mx + b \rightarrow y = 75x + 100$$

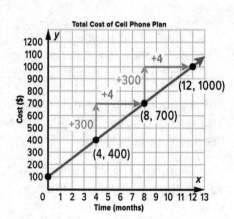

Total Cost of Cell Phone Plan

Think About It!

What is true about the rate of change of a linear function?

Example 1 Construct Linear Functions From Graphs

A shoe store offers free points when you sign up for their rewards card. Then, for each pair of shoes purchased, you earn an additional number of points. The graph shows the total points earned for several pairs of shoes.

Find and interpret the rate of change and initial value. Then write the equation of the function in the form $y = mx + b$.

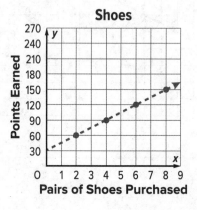

Shoes

Part A Find and interpret the rate of change.

Choose any two points from the graph.

$$\frac{\text{change in points}}{\text{change in pairs}} = \frac{(90 - 60) \text{ points}}{(4 - 2) \text{ pairs}}$$

Use the points (2, 60) and (4, 90).

$$= \frac{30 \text{ points}}{2 \text{ pairs}} \text{ or } \frac{15 \text{ points}}{1 \text{ pairs}}$$

Simplify.

The rate of change is _____, so the number of points earned per pair of shoes is 15.

(continued on next page)

Part B Find and interpret the initial value.

The initial value is called the *y*-intercept. Determine where the line intersects the *y*-axis.

When $x = 0$, $y =$ _____. So, the initial number of points earned is 30.

Part C Write the equation in the form $y = mx + b$.

The rate of change *m*, is 15 and the initial value *b*, is 30.

So, the equation in $y = mx + b$ form is $y = \boxed{} x + \boxed{}$.

Talk About It!

How can you verify that the equation $y = 15x + 30$ correctly represents the function?

Check

A conference center charges an initial deposit plus an hourly rate for renting the space for an event or party. The total cost for different numbers of hours, including the deposit, is shown in the graph.

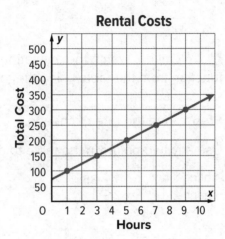

Rental Costs

Part A

Find and interpret the rate of change.

Part B

Find and interpret the initial value.

Part C

Write the equation of the line in the form $y = mx + b$.

Show your work here

Go Online You can complete an Extra Example online.

Learn Construct Linear Functions from Tables

To construct a linear function from a table of values, you need to determine the rate of change and initial value. Then write the equation of the function in the form $y = mx + b$.

The table shows the relationship between the cost of an amusement park membership and the number of months.

Months, x	Cost ($), y
1	35
2	50
3	65
4	80

🞂 **Go Online** Watch the animation to see how to construct a linear function from the table using the following steps.

Step 1 Find the rate of change.

Use any two points in the table to determine the rate of change.

$$\frac{\text{change in } y}{\text{change in } x} = \frac{50 - 35}{2 - 1} \qquad \text{Use the points (1, 35) and (2, 50.)}$$

$$= \frac{15}{1} \text{ or } \boxed{} \qquad \text{Simplify.}$$

Step 2 Find the initial value.

Use the slope-intercept form of a linear equation to find the y-intercept.

$y = mx + b$	Slope-intercept form
$y = 15x + b$	Replace m with the slope, 15.
$35 = 15(1) + b$	Use the point (1, 35). $x = 1, y = 35$
$35 = 15 + b$	Simplify.
$-15 = -15$	Subtract 15 from each side.
$\boxed{} = b$	Simplify.

Step 3 Write the equation of the function.

Substitute the rate of change for m and the initial value for b.

$$y = mx + b \rightarrow y = \boxed{} x + \boxed{}$$

Example 2 Construct Linear Functions From Tables

The table shows how much money Ava has saved. Assume the relationship between the two quantities is linear.

Find and interpret the rate of change and initial value. Then write the equation of the function in the form $y = mx + b$.

Number of Months, x	Money Saved ($), y
3	110
4	130
5	150
6	170

Think About It!

Does the table currently show the initial value? Explain.

Part A Find and interpret the rate of change.

You can use any two points to determine the rate of change.

$$\frac{\text{change in } y}{\text{change in } x} = \frac{130 - 110}{4 - 3} \qquad \text{Use the points (3, 110) and (4, 130).}$$

$$= \frac{20}{1} \text{ or } \boxed{} \qquad \text{Simplify.}$$

The rate of change is 20. So, Ava saves \$_____ each month.

Part B Find and interpret the initial value.

Since the value for y when $x = 0$ is not listed in the table, use the slope-intercept form of a linear equation to find the y-intercept. Substitute values for m, x, and y. You can use any ordered pairs from the table to substitute values for x and y.

$y = mx + b$ Slope-intercept form

$y = \boxed{} x + b$ Replace m with the rate of change, 20.

$\boxed{} = 20\left(\boxed{}\right) + b$ Use the point (3, 110). $x = 3$, $y = 110$

$110 = \boxed{} + b$ Simplify.

$\boxed{} = b$ Subtract 60 from each side.

The y-intercept is 50, so Ava initially saved \$_____.

Part C Write the equation in the form $y = mx + b$.

The rate of change m, is 20 and the initial value b, is 50.

$y = \boxed{} x + \boxed{}$

Check

The table shows the cost of a smartphone plan, based on the data used per month. Assume the relationship between the two quantities is linear.

Number of Gigabytes of Data, x	Cost ($), y
2	50
3	65
4	80
5	95

Part A

Find and interpret the rate of change.

Show your work here

Part B

Find and interpret the initial value.

Show your work here

Part C

Write the equation in the form $y = mx + b$.

🔵 **Go Online** You can complete an Extra Example online.

Pause and Reflect

Did you make any errors when completing the Check exercise? What can you do to make sure you don't repeat that error in the future?

Record your observations here

Learn Construct Linear Functions from Verbal Descriptions

You can write the equation of a function in the form $y = mx + b$ by determining the rate of change and an initial value from a description of the relationship. The rate of change is m, the slope. The initial value is b, the y-intercept.

Go Online Watch the animation to see how to construct a linear function for the following real-world situation.

A skiing instructor charges an initial fee plus $30 per hour for lessons. Tasha paid $205 for six hours of instruction.

Step 1 Find the rate of change.

Since the ski instructor charges $☐ per hour for lessons, the rate of change is 30.

Step 2 Find the initial value.

Use the slope-intercept form of a linear equation to find the y-intercept.

$y = mx + b$	Slope-intercept form
$y = 30x + b$	Replace m with the rate of change, 30.
$205 = 30(6) + b$	Replace y with 205 and x with 6.
$205 = 180 + b$	Simplify.
$\boxed{} = b$	Solve for b.

Step 3 Write the equation of the function.

Substitute the rate of change for m and the initial value for b.

$$y = mx + b \rightarrow y = \boxed{}\,x + \boxed{}$$

Think About It!

What aspect of this scenario describes the rate of change?

Talk About It!

How would the equation of the function be altered if Joan planned to add 8 photos to her photo album each week? Assume that the number of photos at 8 weeks remains 120 photos.

🐦 **Go Online** You can complete an Extra Example online.

🌐 **Example 3** Construct Linear Functions From Verbal Descriptions

Joan plans to add 12 photos to her photo album each week. After 8 weeks, there are 120 photos in the album. Assume the relationship is linear.

Find and interpret the rate of change and initial value. Then write the equation of the function in the form $y = mx + b$, where x represents the number of weeks and y represents the total number of photos in the album.

Part A Find and interpret the rate of change.

Since Joan adds 12 photos to her photo album each week, the rate of change is _____.

Part B Find and interpret the initial value.

Substitute values from the problem for m, x, and y.

$y = mx + b$ Slope-intercept form

$y = \boxed{} x + b$ Replace m with the rate of change, 12.

$\boxed{} = 12 \left(\boxed{} \right) + b$ Replace y with 120 and x with 8.

$120 = \boxed{} + b$ Simplify.

$\boxed{} = b$ Solve for b.

The initial value is 24, so the initial number of photos Joan has is _____.

Part C Write the equation in the form $y = mx + b$.

$y = \boxed{} x + \boxed{}$

Check

Julie is tracking the growth of a plant for a science project. The height of the plant on the 2nd day she measured was 8 inches and on the 7th day it was 20.5 inches. Assume the relationship is linear.

Part A Find and interpret the rate of change.

Part B Find and interpret the initial value.

Part C Write the equation in the form $y = mx + b$.

🌐 Apply Depreciation

Rosa wants to buy a motor scooter. The graph shows information about two different scooters. Assuming the pattern continues, what will be the difference in values of the two scooters when they are each 8 years old?

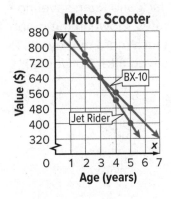

Motor Scooter

1 What is the task?

Make sure you understand exactly what question to answer or problem to solve. You may want to read the problem three times. Discuss these questions with a partner.

First Time Describe the context of the problem, in your own words.
Second Time What mathematics do you see in the problem?
Third Time What are you wondering about?

2 How can you approach the task? What strategies can you use?

Record your observations here

💬 **Talk About It!**
How can you solve this problem another way?

3 What is your solution?

Use your strategy to solve the problem.

Show your work here

4 How can you show your solution is reasonable?

⚡ **Write About It!** Write an argument that can be used to defend your solution.

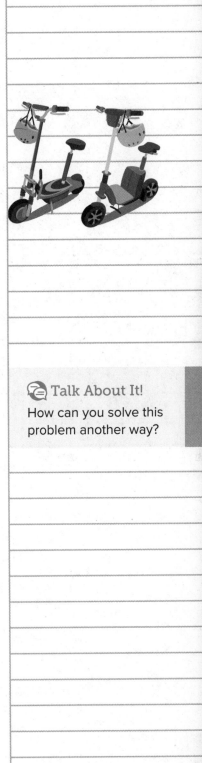

Check

Jack and Ryan save money at the rates shown in the graph. How much more money will Ryan have saved after 20 weeks?

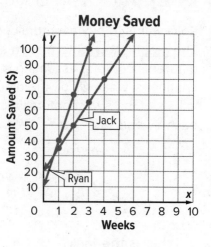

Money Saved

Go Online You can complete an Extra Example online.

Pause and Reflect

Compare and contrast each of the methods used to construct linear functions: from a graph, from a table, from a verbal description.

Practice

Go Online You can complete your homework online.

1. A cleaning service charges an initial fee plus an hourly rate. The total cost for different numbers of hours, including the initial fee, is shown on the graph. Find and interpret the rate of change and initial value. Then write the equation of the function in the form $y = mx + b$. (Example 1)

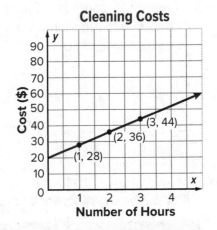

Cleaning Costs

2. The table shows the distance Penelope is from the park as she walks to soccer practice. Assume the relationship between the two quantities is linear. Find and interpret the rate of change and initial value. Then write the equation of the function in the form $y = mx + b$. (Example 2)

Time (min), x	Distance (m), y
5	1,930
10	1,380
15	830
20	280

3. A roller skating rink charges a skate rental fee and an hourly rate to skate. The total cost to skate for 2 hours is $9.50 and for 5 hours is $18.50. Assume the relationship is linear. Find and interpret the rate of change and initial value. Then write the equation of the function in the form $y = mx + b$, where x represents the number of hours and y represents the total cost. (Example 3)

Test Practice

4. **Open Response** A movie theater offers a reward program that charges a yearly membership fee and a discounted rate per movie ticket. The total cost for a reward program member to see 5 movies is $40 and the total cost for 12 movies is $75. Assume the relationship is linear. Write the equation of the function in the form $y = mx + b$, where x represents the number of movies and y represents the total cost.

Apply

5. The graph shows information about the value of two houses in different locations. Assuming the pattern continues, what will be the difference in values of the two houses when they are each 15 years old?

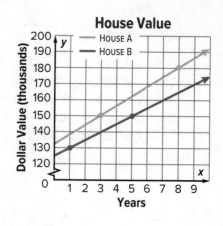

6. The tables show the monthly costs for two different Internet service providers. What will be the difference in the cost of the two plans after 18 months?

Company A	
Month, x	Total Cost ($), y
1	75
2	110
3	145
4	180

Company B	
Month, x	Total Cost ($), y
1	60
2	110
3	160
4	210

7. (MP) **Identify Structure** Determine the rate of change for a horizontal line. Explain why this rate of change makes sense.

8. (MP) **Find the Error** A recreation center charges $50 to rent the gym for 1 hour, and $100 to rent the gym for 3 hours. Jon says the price per hour is $50. Find and correct Jon's mistake. Assume the relationship is linear.

9. Your class is given a real-world scenario of a linear function in which the rate of change is −3 and the initial value is −10. You wrote the equation $y = -3x + (-10)$ to represent the function. Your classmate wrote $y = -3x - 10$. Who is correct? Justify your response.

10. **Create** Write a real-world problem in which you would need to find the rate of change of a linear function. Explain why it would be useful to know the rate of change.

Compare Functions

I Can... compare functions that are represented in different ways using their initial values and rates of change.

Explore Compare Properties of Functions

Online Activity You will use Web Sketchpad to explore how to compare functions presented in different forms.

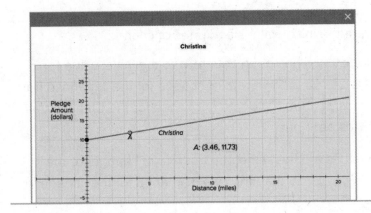

Learn Compare Functions

Functions can be represented by a table, graph, equation, or words. You can compare two functions represented in different forms by comparing the rates of change and initial values. Functions can be compared to determine which is increasing or decreasing at a faster rate, or to determine which function has a greater value for a given input.

The function represented by the table has rate of change of 4, and the function represented by the graph has a rate of change of $\frac{1}{2}$. Therefore, the function represented by the table has a greater rate of change.

Input, x	Output, y
1	4
2	8
3	12
4	16

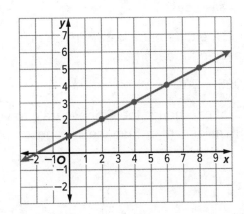

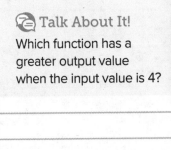

Talk About It!

Which function has a greater output value when the input value is 4?

Think About It!

Without solving the problem, how do you know that the Chinese train travels at a faster rate than the Japanese train?

🌐 Example 1 Compare Two Functions

The function $m = 140h$, where m is the miles traveled in h hours, represents the distance traveled by the first Japanese high-speed train. The distance traveled by a high-speed train operating in China is shown in the table. Assume each relationship is linear.

Chinese High-Speed Train	
Hours	Miles
1	217
2	434
3	651

Compare the functions' initial values and rates of change. Then determine how much farther the Chinese train will travel than the Japanese train, if you ride each train for 5 hours.

Part A Compare the initial values and rates of change.

At 0 hours, no distance has been covered, so the initial value for each function is _____. The rate of change for the Japanese train is _____ miles per hour. The rate of change for the Chinese train is _____ miles per hour.

So, the initial values are the same, but the rate of change for the Chinese train is _____ than the rate of change for the Japanese train.

Part B Determine how much farther the Chinese train will travel than the Japanese train.

Find the distance of the Japanese train by substituting 5 for h into the function.

$m = 140h$	Write the function.
$m = 140(5)$	Replace h with 5.
$m = 700$	Simplify.

Find the distance of the Chinese train by extending the table to 5 hours. The hours increase by 1 each time, and the miles increase by 217.

Find the difference between the miles traveled for each train.

$1,085 - 700 =$ ☐

So, after 5 hours, the Chinese train travels 385 miles farther than the Japanese train.

Chinese High-Speed Train	
Hours	Miles
1	217
2	434
3	651
4	
5	

Check

Suppose you want to decide which movie streaming service you should join. Company A charges $1.50 per movie with a start-up fee of $5. The cost to stream movies from Company B is shown in the table.

Number of Movies, x	Total Cost ($), y
2	8
4	12
6	16

Part A Compare the initial values and rates of change.

Part B How much more will Company B cost than Company A if 15 movies are rented in one month?

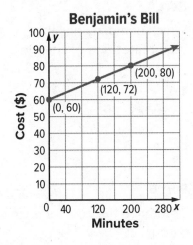

Go Online You can complete an Extra Example online.

🌐 **Example 2** Compare Two Functions

Angela and Benjamin each have a monthly cell phone bill. Angela's monthly cell phone bill costs $0.15 per minute plus an initial fee of $49. Benjamin's monthly cost is shown in the graph.

Compare the functions' initial values and rates of change. Then determine the monthly cost for Angela and Benjamin for 200 minutes.

Part A Compare the initial values and rates of change.

What is the initial value of the function for Angela's bill? _____

What is the initial value for Benjamin's bill? _____

So, Benjamin has a greater initial cost.

What is the rate of change for Angela's bill? _____

What is the rate of change for Benjamin's bill? _____

So, Angela pays more per minute.

> **Think About It!**
> How can you determine the initial value for Benjamin's cell phone plan?

(continued on next page)

Talk About It!

At 200 minutes, who has to pay more? Will this person always have to pay more each month? Explain.

Part B Determine the monthly cost for Angela and Benjamin for 200 minutes.

Use the values given in the problem to find Angela's cost.
At 200 minutes, Angela will pay 0.15(200) + 49, or $_____.

Use the graph to find Benjamin's cost.
At 200 minutes, Benjamin will pay $_____.

Check

You and your friends want to go on a canoe trip. River Run Canoes charges an initial fee of $10 plus $10 per hour. The cost for Clear Water Canoes is represented by the graph.

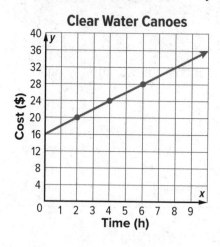

Clear Water Canoes

Part A

Which company has the greater initial value and which company has the greater rate of change?

Part B

Determine the cost for each company if you want to canoe for 4 hours.

Go Online You can complete an Extra Example online.

Apply eBooks

Aisha is researching three different online book subscription companies. The cost y of each subscription is a linear function of the number of books downloaded, x. The prices for three different subscription companies are shown. Aisha plans to download 10 books and wants to spend the lesser amount. From which company should she purchase a subscription?

The Bookshelf: $y = 4.5x + 20$

We-Reads

Number of Books	Cost ($)
0	18
1	23.75
2	29.50
3	35.25
4	41

Bookworm

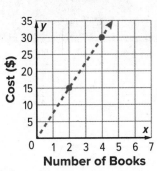

1 What is the task?

Make sure you understand exactly what question to answer or problem to solve. You may want to read the problem three times. Discuss these questions with a partner.

First Time Describe the context of the problem, in your own words.
Second Time What mathematics do you see in the problem?
Third Time What are you wondering about?

2 How can you approach the task? What strategies can you use?

3 What is your solution?

Use your strategy to solve the problem.

4 How can you show your solution is reasonable?

✐ **Write About It!** Write an argument that can be used to defend your solution.

Talk About It!

How could you solve the problem another way?

Check

Andrea is researching cell phone plans to determine which is the least expensive. The cost y of each plan is a linear function of the number of months x that she has the plan. The prices for three different plans are shown. Andrea expects to stay on the plan for 18 months and wants to spend the lesser amount. Which plan should she choose?

Cell Phone Plan A: $y = 50x + 50$

Cell Phone Plan B

Number of Months	Cost ($)
0	100
1	145
2	190
3	235

Cell Phone Plan C

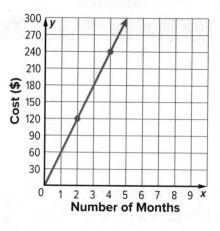

Go Online You can complete an Extra Example online.

Pause and Reflect

How do you feel when you are asked during class to answer a question or to explain a solution?

Record your observations here

Practice

🅖 Go Online You can complete your homework online.

1. Gennaro is considering two job offers as a part-time sales person. Company A will pay him $12.50 for each item he sells, plus a base salary of $500 at the end of the month. The amount Company B will pay him at the end of the month is shown in the table. Compare the functions' initial values and rates of change. Then determine how much more Gennaro would make at Company A if he sells 28 items by the end of the month. (Example 1)

Number of Items Sold, x	Total Earned ($), y
5	425
10	500
15	575

2. The temperature in two different ovens increased at a steady rate. The temperature in oven A is represented by the equation $y = 25x + 72$, where x represents the number of minutes and y represents the temperature in degrees Fahrenheit. The temperature of oven B is shown in the graph. Compare the functions' initial values and rates of change. Then determine how much greater the temperature in oven B will be than oven A after 8 minutes. (Example 1)

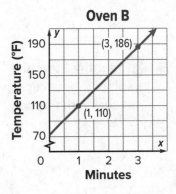

Oven B

Test Practice

3. **Open Response** Lorraine and Chila were riding their bikes to school. Lorraine's distance away from the school is shown in the graph. Chila's distance away from the school is shown in the table. Compare the functions' initial values and rates of change. Then determine Lorraine's and Chila's distance from school after 7 minutes. (Example 2)

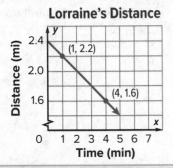

Lorraine's Distance

Chila's Distance	
Time (min), x	Distance (mi), y
1	1.5
2	1.3
3	1.1

Apply

4. Samuel is planning a walking tour of his city and is researching three different tour-guide companies. The cost y of each company is a linear function of the number of hours x that someone spends on the tour. The prices for three different companies are shown. Samuel plans his tour for 6 hours and wants to spend the lesser amount of money. From which company should he choose to do his walking tour?

Hidden Treasures Walking Tour
$y = 9x + 25$

Road-Less-Traveled Tours	
Hours, x	Cost ($), y
1	12
2	24

Bowman Bros.

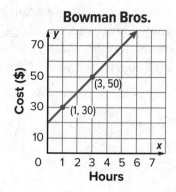

5. **MP Identify Structure** Explain why the graph of the function $y = 7x + 3$ will never intersect with the graph of the function $y = 7x + 10$.

6. **MP Justify Conclusions** Tammy was comparing information for two cell phone plans. Plan A would cost $480 for one year. Plan B would cost $250 for 6 months. Tammy reasons that plan B must cost more for one year. Is there enough information to know if she is correct? Explain your reasoning.

7. Indicate whether the following statement is *always*, *sometimes*, or *never* true. Justify your reasoning.

When two functions have different rates of change, and different initial values, the function with the greater rate of change will have the greater output value for every input value.

8. **Create** Write a real-world problem in which you would need to compare initial values for three linear functions.

Nonlinear Functions

I Can... determine if a function, represented in different forms, is a linear or nonlinear function by using the rate of change, shape of the graph, or structure of the equation.

What Vocabulary Will You Learn?
nonlinear functions

Explore Linear and Nonlinear Functions

Online Activity You will use Web Sketchpad to explore how to determine if a function is linear or nonlinear.

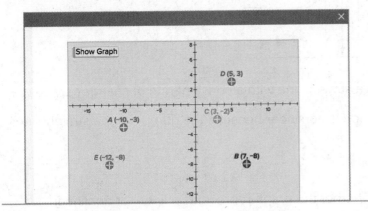

Learn Identify Linear and Nonlinear Functions from Graphs

Linear functions have graphs that are straight lines, because the rate of change between any two data points is constant. **Nonlinear functions** are functions whose rates of change are not constant, and therefore, their graphs are not straight lines.

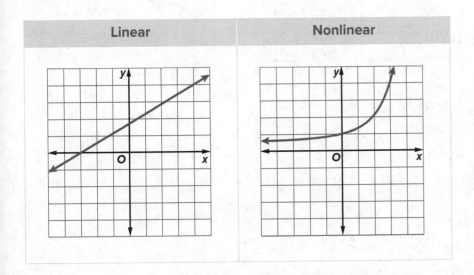

Linear	Nonlinear

💬 **Talk About It!**

The graph is composed of multiple straight lines. Why is it nonlinear?

Example 1 Identify Linear and Nonlinear Functions from Graphs

Determine whether the graph represents a linear or nonlinear function. Explain.

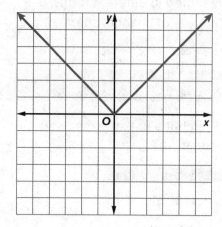

Is the graph a straight line with a constant rate of change? _____

So, the function is nonlinear because the graph is not a straight line.

Check

Determine whether the graph represents a linear or nonlinear function. Explain.

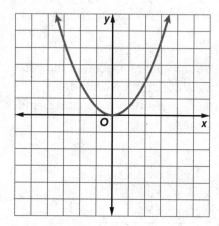

Show your work here

🔵 **Go Online** You can complete an Extra Example online.

Example 2 Identify Linear and Nonlinear Functions from Graphs

A square has a side length of s inches. The area of the square is a function of the side length.

Graph the function to determine if it is linear or nonlinear. Explain.

Think About It!

What do you need to do before graphing the function?

Step 1 Make a table of values.

Complete the table to show the area of the square for side lengths of 1, 2, 3, 4, and 5 inches. The formula for the area of a square is $A = s^2$.

Side length (in.)	1	2	3	4	5
Area (in^2)					

Step 2 Graph the function.

Graph the values in the table as ordered pairs, where x is the side length and y is the area. Then connect the points with a smooth curve.

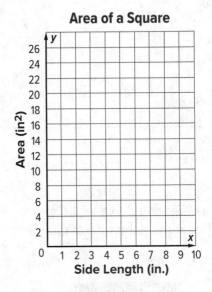

Area of a Square

Do the points lie on a straight line? _____

So, the graph is nonlinear.

Check

A cube has a side length of s meters. The volume of the cube is represented by the expression s^3. The volume of a cube is a function of the side length. Does this situation represent a linear or nonlinear function? Explain.

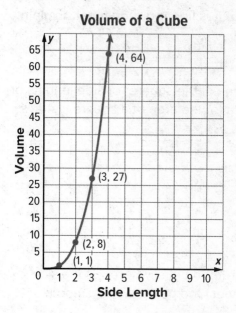

Volume of a Cube

Show your work here

🧭 **Go Online** You can complete an Extra Example online.

Learn Identify Linear and Nonlinear Functions from Tables

You can determine whether a function represented in a table is linear or nonlinear. To do so, find the rates of change. If the rate of change is constant, the relationship is linear.

Linear	
x	y
5	8
10	6
15	4
20	2

+5, +5, +5 / −2, −2, −2

Nonlinear	
x	y
1	1
2	4
3	9
4	16

+1, +1, +1 / +3, +5, +7

The rate of change is always $-\frac{2}{5}$.

So, the function is _____.

The rates of change, $\frac{3}{1}$, $\frac{5}{1}$ and $\frac{7}{1}$, are not the same. So, the

function is _____.

Example 3 Identify Linear and Nonlinear Functions from Tables

The table shows the minimum number of daily Calories a tiger cub should eat based on its age in weeks.

Age (weeks)	Minimum Calorie Intake
1	825
2	1,000
3	1,185
4	1,320
5	1,420

Determine whether the function is linear or nonlinear. Explain.

From Week 1 to Week 2, the rate of change is _____.

From Week 2 to Week 3, the rate of change is _____.

The rates of change are not the same. So, the function is nonlinear.

Check

Determine whether the table represents a linear or nonlinear function. Explain.

Time, x	Distance, y
3	9
4	16
5	25
6	36

Show your work here

Go Online You can complete an Extra Example online.

Think About It!

What are the characteristics of a nonlinear function?

Talk About It!

If the rates of change were the same between Weeks 1 and 2, and between Weeks 2 and 3, would that be enough information to determine that the function is linear? Explain.

Learn Identify Linear and Nonlinear Functions from Equations

In order for an equation to represent a linear function, the equation must be able to be written in the form $y = mx + b$, where m represents the slope or constant rate of change and b represents the y-intercept or initial value.

If an equation represents a nonlinear function, it will have one or more of the following characteristics.

- powers other than one on the variable
- square or cube roots of variable expressions
- variables in the denominator of a fraction

Go Online Watch the animation to see the following examples of different types of nonlinear equations.

$y = x^2 + 3$

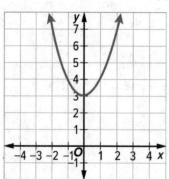

$y = 4x^3$

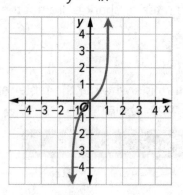

$y = \sqrt{2x + 3}$

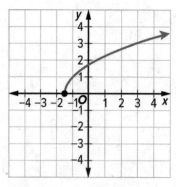

$y = \frac{2}{x}$

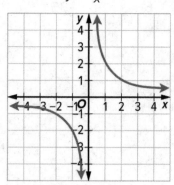

Example 4 Identify Linear and Nonlinear Functions from Equations

Determine whether the equation represents a linear or nonlinear function. Explain.

$y = \frac{x}{3}$

The equation $y = \frac{x}{3}$ can also be written as $y = \frac{1}{3}x$. Is the equation $y = \frac{1}{3}x$ in the form $y = mx + b$? _____

So, the equation $y = \frac{x}{3}$ represents a linear function because it can be written in the form $y = mx + b$, where m is $\frac{1}{3}$ and b is 0.

Check

Determine whether the equation represents a linear or nonlinear function. Explain.

$y = \frac{-2x}{5} + 5$

Example 5 Identify Linear and Nonlinear Functions from Equations

Determine whether the equation represents a linear or nonlinear function. Explain.

$y = 3x^2 + 2x + 5$

The equation $y = 3x^2 + 2x + 5$ cannot be written in the form $y = mx + b$ because the variable x is raised to a power other than 1.

So, the equation $y = 3x^2 + 2x + 5$ is nonlinear because it _____ be written in the form $y = mx + b$.

Check

Determine whether the equation represents a linear or nonlinear function. Explain.

$y = \sqrt{3x + 5}$

 Go Online You can complete an Extra Example online.

Pause and Reflect

Create a graphic organizer that will help you study the concepts you learned today in class.

Record your observations here

Math History Minute

Amalie "Emmy" Noether (1882–1935) was a German mathematician. As a pioneering female mathematician, Noether helped explore the mathematics behind Albert Einstein's theory of general relativity. She also discovered a mathematical relationship that applies to physics, now known as Noether's theorem.

Apply Geometry

The circumference of a circle is a function of the circle's radius. Similarly, the area of a circle is also a function of the circle's radius. Which of these relationships is a linear function? Construct an argument to justify your response.

1 What is the task?

Make sure you understand exactly what question to answer or problem to solve. You may want to read the problem three times. Discuss these questions with a partner.

First Time Describe the context of the problem, in your own words.
Second Time What mathematics do you see in the problem?
Third Time What are you wondering about?

2 How can you approach the task? What strategies can you use?

Record your observations here

3 What is your solution?

Use your strategy to solve the problem.

Show your work here

Talk About It!
How can you solve the problem another way?

4 How can you show your solution is reasonable?

Write About It! Write an argument that can be used to defend your solution.

Check

The area of a square is a function of its perimeter. Similarly, the perimeter of a square is a function of its side length. Which of these relationships is a linear function? Construct an argument to justify your response.

Go Online You can complete an Extra Example online.

Foldables It's time to update your Foldable, located in the Module Review, based on what you learned in this lesson. If you haven't already assembled your Foldable, you can find the instructions on page FL1.

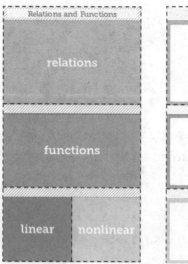

Practice

Go Online You can complete your homework online.

Determine whether each graph represents a linear or nonlinear function. Explain. (Example 1)

1.

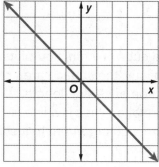

2.

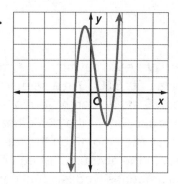

3. A rectangle has side lengths s and $2s$. Its area is represented by the expression $2s^2$. The area of the rectangle is a function of its side length. Does this situation represent a linear or nonlinear function? Explain. (Example 2)

Area of a Rectangle

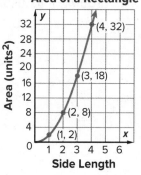

Determine whether each table represents a linear or nonlinear function. Explain. (Example 3)

4.

Number of Items Sold, x	Total Earned ($), y
1	25
2	45
3	60

5.

Time (min), x	Distance (mi), y
15	2.2
30	4.4
45	6.6

Determine whether each equation represents a linear or nonlinear function. Explain. (Examples 4 and 5)

6. $y + 7x = 2$

7. $y = \sqrt{8x}$

8. **Multiselect** Select all equations that represent a linear function.

☐ $y = 7^2$ ☐ $y = \frac{-4}{x}$ ☐ $y + \frac{1}{2}x = 9$

☐ $9 = y - \sqrt[3]{2x}$ ☐ $y = x^2$ ☐ $2.3 = y - 5.8x$

Apply

9. Catalina and Terri are each saving money to buy a car. Catalina puts $1,000 in a safe at home and adds $25 every month. Terri puts $500 in a savings account at the bank that earns 3.2% interest each month on the total amount of money in the account. For each person, the total amount saved is a function of the number of months. Which of these relationships is a linear function? Construct an argument to justify your response.

10. The volume of a cube is a function of its side length. Similarly, the surface area of a cube is a function of its side length. Which of these relationships is a linear function? Construct an argument to justify your response.

11. ⓂⓅ **Find the Error** Semih says the graph shown is a linear function because the graph lies in a straight line. Correct the error.

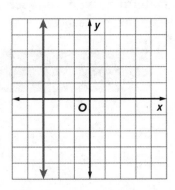

12. ⓂⓅ **Identify Structure** Write the equations for two different functions that are nonlinear. Then explain how the equations are different than the equations of linear functions.

13. Determine if the following statement is *true* or *false*. Explain your reasoning.

   The graph of a non-vertical straight line is always a function, but the graph of a function is not always a straight line.

Qualitative Graphs

I Can... recognize a qualitative graph and interpret the scenario it represents as well as create a qualitative graph.

What Vocabulary Will You Learn?
qualitative graphs

Explore Interpret Qualitative Graphs

Online Activity You will use Web Sketchpad to explore how to interpret qualitative graphs.

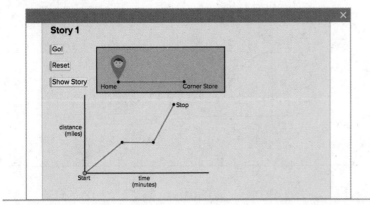

Learn Analyze Qualitative Graphs

The graph shown is a qualitative graph. **Qualitative graphs** are graphs used to represent situations that may not have numerical values, or graphs in which numerical values are not included.

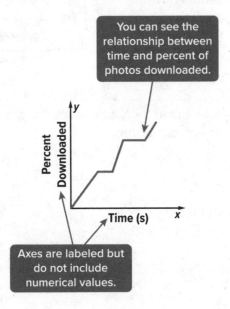

You can see the relationship between time and percent of photos downloaded.

Axes are labeled but do not include numerical values.

(continued on next page)

The rate of change of a qualitative graph can provide useful information between two values.

increasing at a
constant rate

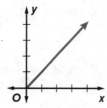

decreasing at a
constant rate

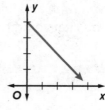

no change

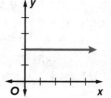

increasing at a
varied rate

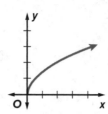

decreasing at a
varied rate

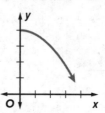

🌐 Example 1 Analyze Qualitative Graphs

The graph displays the water level in a bathtub. Describe the change in the water level over time.

The graph demonstrates how the water level changes as time increases.

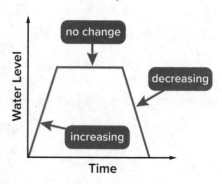

So, the water level _____ at a constant rate. For a period of time, there is no change in the water level. Then, the water level _____ at a constant rate.

Check

The graph represents revenue from a local clothing store. Describe the sales over time.

🧭 **Go Online** You can complete an Extra Example online.

Think About It!

How will the different rates of change help you describe the changes in the water level?

Talk About It!

Describe a real-world scenario that would result in the water level of a bathtub changing as shown in the graph.

Learn Sketch Qualitative Graphs

Qualitative graphs represent essential elements of a situation in a graphical form. You can sketch qualitative graphs to represent many real-world functions that are described verbally.

Go Online Watch the video to learn how to sketch qualitative graphs for the following scenarios.

Charlie's speed over time:

Charlie is riding his bike. He speeds up, rides at a constant rate, then slows down.

Sketch a qualitative graph to represent the scenario.

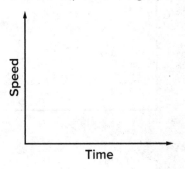

Number of pencils sold over time:

A school store sells pencils. At the beginning of the year, they sell a lot of pencils, then sales drop, and finally, pencils are sold at a steady rate.

Sketch a qualitative graph to represent the scenario.

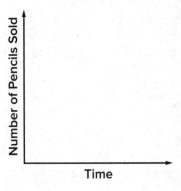

🌐 Example 2 Sketch Qualitative Graphs

A tennis ball is dropped on the floor. On each successive bounce, it rebounds to a height less than its previous bounce height until it comes to rest on the floor.

Sketch a qualitative graph to represent the situation. Determine if the graph is linear or nonlinear and where the graph is increasing or decreasing.

Part A Sketch a qualitative graph to represent the situation.

Draw and label the axes. Label the vertical axis "Distance from Floor." Label the horizontal axis "Time."

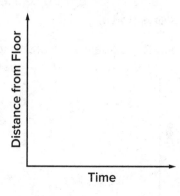

Sketch the shape of the graph. The distance from the floor starts out at a high value, then falls to the floor. The ball bounces and rebounds to a height less than its drop height. This pattern is repeated several times until the ball comes to rest on the floor.

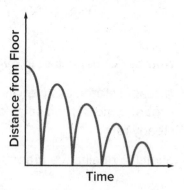

Part B Determine if the graph is linear or nonlinear and where the graph is increasing or decreasing.

Is the rate of change in the graph constant? _____

So, the graph is nonlinear.

The graph is _____ whenever the tennis ball travels up in the air, because the distance from the floor is increasing. The graph is _____ whenever the tennis ball travels towards the ground, because the distance from the floor is decreasing.

💭 **Think About It!**

What label would you plot along the horizontal axis? the vertical axis?

Check

A person riding a bike slowly decreases their speed. Then they quickly speed up. Finally, they maintain a constant speed for a period of time.

Part A

Sketch a qualitative graph to represent the situation.

Part B

Determine if the graph is linear or nonlinear and where the graph is increasing or decreasing.

Show your work here

 Go Online You can complete an Extra Example online.

Pause and Reflect

How does what you already know about rate of change help you with sketching qualitative graphs?

Record your observations here

Luis is biking to his friend Nathan's house. He increases his speed at a constant rate for 2 minutes until he reaches a speed of 6 miles per hour. He then decreases his speed at a constant rate for a minute until he reaches a speed of 2 miles per hour. Finally, his speed increases at a varied rate for 3 minutes until he reaches a speed of 8 miles per hour.

Sketch a graph that accurately represents Luis's trip.

Draw and label the axes. Label the vertical axis "Speed (mph)." Label the horizontal axis "Time (min)."

Use the information in the problem to sketch the shape of the graph.

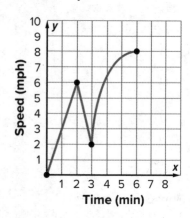

Check

Julian is riding his horse. The horse's speed increases at a constant rate for 2 minutes until it reaches a speed of 10 miles per hour. Then, it increases his speed at a varied rate for 1 minute until it reaches a speed of 20 miles per hour. Finally, the horse decreases its speed at a constant rate for 3 minutes until it comes to a stop. Sketch a graph that accurately represents Julians' horse ride.

Show your work here

🐦 **Go Online** You can complete an Extra Example online.

Think About It!

How would you begin sketching the graph?

Practice

Go Online You can complete your homework online.

1. The graph displays the distance Wesley was from home as he ran in preparation for his cross-country meet. Describe the change in distance over time. (Example 1)

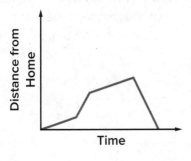

2. The graph displays the population of bacteria in a petri dish. Describe the change in population over time. (Example 1)

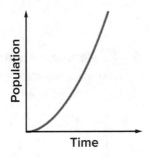

3. Ryan's heart rate was steady before exercising. While exercising, his heart rate increased rapidly and then steadied. During cool down, his heart rate decreased slowly then lowered quickly until becoming steady again. Sketch a qualitative graph to represent the situation. Determine if the graph is linear or nonlinear and where the graph is increasing or decreasing. (Example 2)

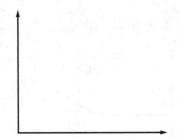

4. An oven is being preheated. The temperature starts at 75°F and increases at a constant rate for 8 minutes until it reaches the desired temperature, 350°F. It remains the same temperature for 27 minutes. Then the temperature decreases at a constant rate for 5 minutes until it reaches 175°, where it remains steady to keep the food warm. Sketch a graph to represent the situation. (Example 3)

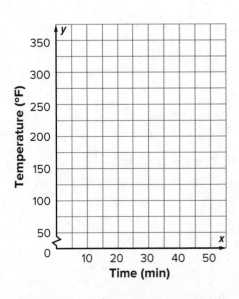

5. **Open Response** A well is being dug on a piece of land. The graph displays the depth of the well over time. Describe the change in the depth of the well over time.

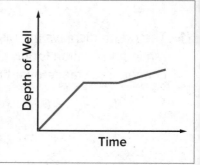

6.  **Persevere with Problems** The graph shows the speed of a train as time increases. Draw a graph and describe how the distance of the train changes as time increases.

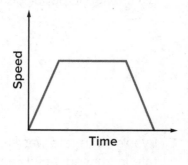

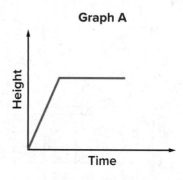

7. A plant grows steadily until it reaches its full height, at which time it stops growing. Which graph displays this relationship? Explain your reasoning.

Graph A   **Graph B**   **Graph C**

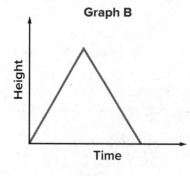 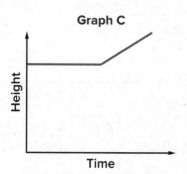

8. **Create** Describe a real-world situation in which a qualitative graph can be used to represent the function.

Tape here

Tab 3	Relations and Functions
Tab 2	
Tab 1	

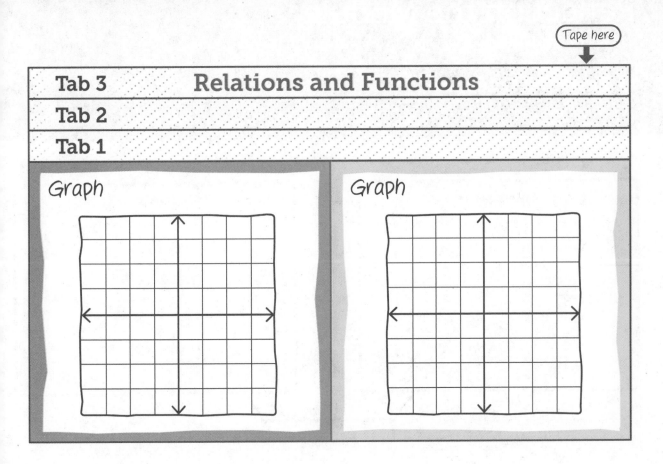

Graph

Graph

Rate Yourself! ⬛ ◈ ★

Complete the chart at the beginning of the module by placing a checkmark in each row that corresponds with how much you know about each topic after completing this module.

Write about one thing you learned.

_____

_____

_____

_____

_____

_____

Write about a question you still have.

_____

_____

_____

_____

_____

_____

Reflect on the Module

Use what you learned about functions to complete the graphic organizer.

e Essential Question

What does it mean for a relationship to be a function?

Ordered Pairs

Table

Functions

Graph

Mapping Diagram

Test Practice

1. Multiselect Which of the following relations are functions? Select all that apply. **(Lesson 1)**

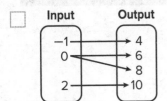

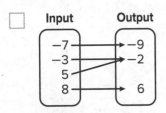

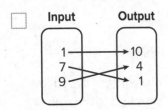

x	−8	−2	0	3
y	1	2	3	4

☐ {(1, 4), (1, 5), (2, 5), (3, 6)}

2. Open Response Determine whether the relation shown in the table is a function. Justify your reasoning. **(Lesson 1)**

Input, x	Output, y
9	2
6	4
9	6
12	8

3. Grid Graph the function $y = -x + 4$, represented by the given function table. **(Lesson 2)**

Input, x	Output, y
−4	8
−2	6
0	4
2	2

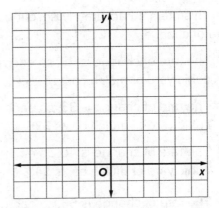

4. Multiple Choice A county park charges $12 per hour to rent a row boat, plus a $20 deposit. The total cost y to rent a boat for x hours can be represented by the function $y = 12x + 20$. **(Lesson 2)**

A. Which set of numbers represents sensible input values for this situation?

Ⓐ integers

Ⓑ decimals

Ⓒ whole numbers

Ⓓ irrational numbers

B. Complete the function table for $y = 12x + 20$, by providing the y values.

Input, x	Output, y
1	
2	
3	
4	

5. Multiple Choice The table shows the total cost for the flat delivery fee and purchase of mulch, based on the number of cubic yards of mulch. Assume the relationship between the two quantities is linear. (Lesson 3)

Number of Cubic Yards, x	Total Cost ($), y
2	110
3	145
4	180
5	215

A. Which statement accurately describes the rate of change?

Ⓐ The rate of change is 25, so each cubic yard of mulch costs $25.

Ⓑ The rate of change is 35, so each cubic yard of mulch costs $35.

Ⓒ The rate of change is 40, so each cubic yard of mulch costs $40.

Ⓓ The rate of change is 55, so each cubic yard of mulch costs $55.

B. Find and interpret the initial value. Then write the equation of the function.

6. Open Response Fitness Center A charges it members $32 per month with a start-up fee of $100. The membership cost of Fitness Center B is shown in the table. Assume each relationship is linear. Compare the functions' initial values and rates of change. (Lesson 4)

Number of Months, x	Total Cost ($), y
2	80
3	120
4	160

7. Table Item Determine whether each equation represents a linear or nonlinear function. (Lesson 5)

	Linear	Nonlinear
$y = 2x - 9$		
$y = \frac{5}{x}$		
$y = 3x^2$		
$4x + y = 7$		
$y = \frac{x}{2}$		
$y = \sqrt{x} + 3$		

8. Multiple Choice Which one of the following descriptions could be used to explain the qualitative graph shown? (Lesson 6)

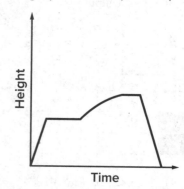

Ⓐ A hiker climbs a cliff at a constant rate for a period of time. Then the hiker continued to climb at a varied rate, before descending at a constant rate.

Ⓑ A hiker climbs a cliff at a constant rate for a period of time. The hiker rests for a period of time. Then the hiker continues to ascend at a varied rate, before resting again and then descending at a constant rate.

Ⓒ A hiker descends a cliff at a constant rate, stops to rest half-way along the way for a period of time, before descending at a constant rate.

Ⓓ A hiker descends a cliff at a varied rate, stops to rest half-way along the way for a period of time, before descending at a constant rate.

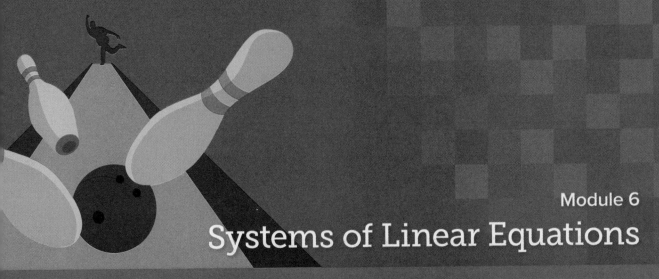

Module 6

Systems of Linear Equations

℮ Essential Question

How can systems of equations be helpful in solving everyday problems?

What Will You Learn?

Place a checkmark (✓) in each row the corresponds with how much you already know about each topic **before** starting this module.

KEY			Before			After		
⬛ — I don't know. ◆ — I've heard of it. ★ — I know it!			⬛	◆	★	⬛	◆	★
solving systems of equations by graphing								
determining the number of solutions to a system of equations								
solving systems of equations by substitution								
solving systems of equations by elimination								
writing and solving systems of equations								

📁 Foldables Cut out the Foldable and tape it to the Module Review at the end of the module. You can use the Foldable throughout the module as you learn about systems of linear equations.

What Vocabulary Will You Learn?

Check the box next to each vocabulary term that you may already know.

□ elimination

□ solution

□ substitution

□ system of equations

Are You Ready?

Study the Quick Review to see if you are ready to start this module.
Then complete the Quick Check.

Quick Review

Example 1

Graph linear equations.

Graph $y = -x + 2$.

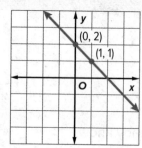

Use the slope, −1, and y-intercept, 2, to graph the equation, or make a table of values.

Example 2

Solve equations with variables on each side.

Solve $2x + 7 = -3x - 13$ for x.

$2x + 7 = -3x - 13$	Write the equation.
$+3x \qquad = +3x$	Addition Property of Equality
$5x + 7 = -13$	Simplify.
$-7 = -7$	Subtraction Property of Equality
$5x = -20$	Simplify.
$\dfrac{5x}{5} = \dfrac{-20}{5}$	Division Property of Equality
$x = -4$	Simplify.

Quick Check

1. The equation $y = 2x + 3$ represents the number of hits, y, a team has after x games. Graph the equation.

Hits per Game

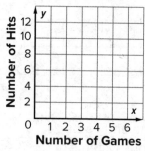

2. The equation $4x - 5 = 2x + 25$ represents the number of points, x, Jack scored in the last game. Solve the equation for x.

How Did You Do?

Which exercises did you answer correctly in the Quick Check?
Shade those exercise numbers at the right.

Solve Systems of Equations by Graphing

I Can... write equations in slope-intercept form in order to graph them and use the graphs to solve a system of equations.

What Vocabulary Will You Learn?
system of equations
solution

Explore Systems of Equations

Online Activity You will use Web Sketchpad to explore what it means when two linear equations intersect and make a conjecture about the point of intersection.

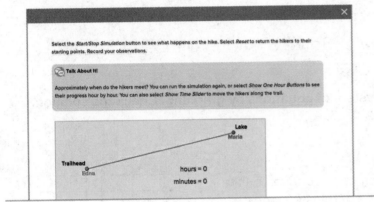

Select the *Start/Stop Simulation* button to see what happens on the hike. Select *Reset* to return the hikers to their starting points. Record your observations.

Talk About It!

Approximately when do the hikers meet? You can run the simulation again, or select *Show One Hour Buttons* to see their progress hour by hour. You can also select *Show Time Slider* to move the hikers along the trail.

Lake
Maria

Trailhead
Edna

hours = 0
minutes = 0

Learn System of Equations

Two or more equations with the same set of variables are called a **system of equations**. For example, $y = 5x$ and $y = 2x + 30$, together, are a system of equations.

The **solution** of this system is (10, 50) because the ordered pair is a solution of both equations.

$y = 5x$ Write the equations. $y = 2x + 30$

$\boxed{} \stackrel{?}{=} 5\left(\boxed{}\right)$ Replace (x, y) with (10, 50). $\boxed{} \stackrel{?}{=} 2\left(\boxed{}\right) + 30$

$\boxed{} = \boxed{}$ Simplify. $\boxed{} = \boxed{}$

The solution of the system is (10, 50). This means that when $x = 10$, the expressions $5x$ and $2x + 30$ both have a value of 50. In other words, $y = 50$ when $x = 10$.

Talk About It!

Do you think (10, 50) is the only solution to this system of equations? Why or why not?

Learn Solve System of Equations by Graphing

You can estimate the solution of a system of equations by graphing the equations on the same coordinate plane. The ordered pair for the point of intersection of the graphs is the solution of the system.

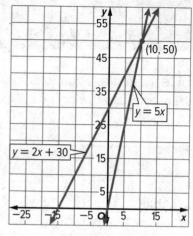

In the graph shown, the lines intersect at only _____ point. The system has _____ solution.

💭 **Think About It!**

Analyze the equations. In which quadrant do you think the lines will intersect? Explain your reasoning.

Example 1 Solve Systems with One Solution by Graphing

Solve the system of equations by graphing. Check the solution.
$$y = -2x - 3$$
$$y = 2x + 5$$

Step 1 Graph each equation on the same coordinate plane.

At what point do the graphs of the lines appear to intersect?

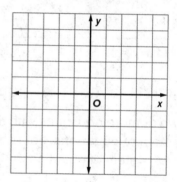

💬 **Talk About It!**

Why did you use algebra to check the solution (−2, 1)?

Step 2 Check the ordered pair.

To verify that the ordered pair is the solution, replace x with −2 and y with 1 in each equation.

$$y = -2x - 3$$
$$1 \overset{?}{=} -2(-2) - 3$$

☐ = ☐

$$y = 2x + 5$$
$$1 \overset{?}{=} 2(-2) + 5$$

☐ = ☐

Since both sentences are true, the solution of the system of equations is (−2, 1).

Check

Solve the system of equations by graphing.

$y = 4x$

$y = x + 3$

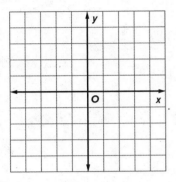

 Go Online You can complete an Extra Example online.

Example 2 Solve Systems with One Solution by Graphing

Solve the system of equations by graphing. Check the solution.

$x = 3$

$y = -6$

Step 1 Graph each equation on the same coordinate plane.

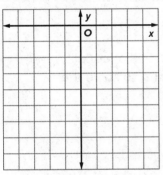

At what point do the graphs of the lines appear to intersect?

Step 2 Check the ordered pair.

Analyze the equations $x = 3$ and $y = -6$ from the system. Since x is equal to _____ and y is equal to _____, the solution of the system is $(3, -6)$.

🕹 **Think About It!**

Without graphing, what kind of line is represented by each equation? Explain how you determined this.

Check

Solve the system of equations by graphing.

$x = -1$

$y = -4$

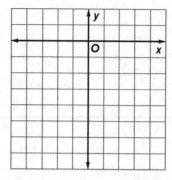

Go Online You can complete an Extra Example online.

Learn Write Linear Equations in Slope-Intercept Form

If a linear equation is not in slope-intercept form, you can use the properties of equality to rewrite the equation.

Go Online Watch the animation to see how to write $4x + 2y = 16$ in slope-intercept form.

The animation shows how you can use properties of equality, and what you already know about solving equations, to solve the equation $4x + 2y = 16$ for y.

$$4x + 2y = 16$$

$$\boxed{} = \boxed{} \qquad \text{Subtraction Property of Equality}$$

$$2y = 16 - 4x \qquad \text{Simplify.}$$

$$\frac{2y}{\boxed{}} = \frac{16 - 4x}{\boxed{}} \qquad \text{Division Property of Equality}$$

$$y = \boxed{} - \boxed{} \qquad \text{Simplify.}$$

$$y = \boxed{} + \boxed{} \qquad \text{Commutative Property}$$

Talk About It!

In the fourth line of the solution, why is it important to divide each term of the equation by 2?

Learn Systems of Equations with No Solution

Some systems of equations have no solution. If the graphs of the lines are parallel, and therefore never intersect, then there is no solution.

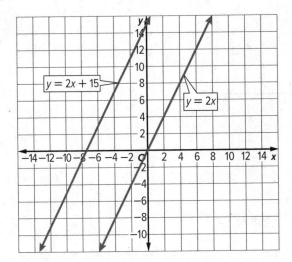

In the graph shown, there is _____ intersection point. The system has _____ solution.

Talk About It!

Use the graph and the structure of the equations to explain why it makes sense that the system has no solution.

Example 3 Solve Systems with No Solution by Graphing

Solve the system of equations by graphing. Check the solution.

$$y = \frac{2}{3}x + 1$$

$$y - \frac{2}{3}x = -3$$

Step 1 Write both equations in slope-intercept form.

The equation $y = \frac{2}{3}x + 1$ is already in slope-intercept form. Write the equation $y - \frac{2}{3}x = -3$ in slope-intercept form.

$y - \frac{2}{3}x = -3$ Write the equation.

$\boxed{} = \boxed{}$ Add $\frac{2}{3}x$ to each side.

$y = -3 + \frac{2}{3}x$ Simplify.

$y = \boxed{}$ Write in slope-intercept form.

Think About It!

What do you need to do first before graphing the equations?

(continued on next page)

Step 2 Graph each equation on the same coordinate plane.

$$y = \frac{2}{3}x + 1$$

$$y = \frac{2}{3}x - 3$$

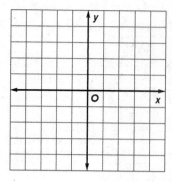

Do the lines appear to intersect?

The graphs of the lines appear to be parallel lines. Since there is no point that is a solution of both equations, there is no solution for this system of equations.

Step 3 Check the solution.

Analyze the equations. Rewrite the equation $y = \frac{2}{3}x + 1$.

$$y = \frac{2}{3}x + 1 \qquad \text{Write the equation.}$$

$$\boxed{} = \boxed{} \qquad \text{Subtract } \frac{2}{3}x \text{ from each side.}$$

$$\boxed{} = 1 \qquad \text{Simplify.}$$

So, the system of equations can be written as follows.

$$y - \frac{2}{3}x = 1$$

$$y - \frac{2}{3}x = -3$$

Since $y - \frac{2}{3}x$ cannot simultaneously be 1 and −3, there is

_____ solution.

Check

Solve the system of equations by graphing.

$y - 2x = 4$

$y = 2x$

Show
your work
here

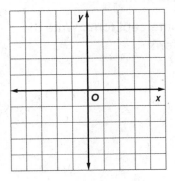

🄽 **Go Online** You can complete an Extra Example online.

Learn Systems of Equations with Infinitely Many Solutions

Some systems of equations have infinitely many solutions. If the graphs of the lines are the same, and therefore intersect at every point, then there is an infinite number of solutions.

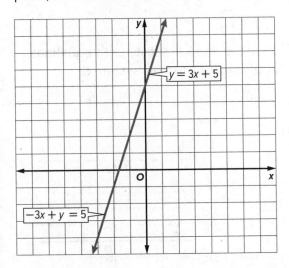

$y = 3x + 5$

$-3x + y = 5$

In the graph shown, the lines intersect at _____ point. The system has an _____ number of solutions.

💬 **Talk About It!**

Use the graph and the structure of the two equations to explain why it makes sense that the system has an infinite number of solutions.

Example 4 Solve Systems by Graphing: Infinitely Many Solutions

Solve the system of equations by graphing.

$y = 2x + 1$

$y - 3 = 2x - 2$

Step 1 Write both equations in slope-intercept form.

The equation $y = 2x + 1$ is already in slope-intercept form. Write the equation $y - 3 = 2x - 2$ in slope-intercept form.

$y - 3 = 2x - 2$	Write the equation.
$\underline{+ 3 = + 3}$	Add 3 to each side.
$y = \boxed{} + \boxed{}$	Simplify.

Both equations are the same.

Step 2 Graph the equations on the same coordinate plane.

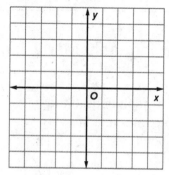

Any ordered pair on the graph of the line will satisfy both equations. So, there is an infinite number of solutions for this system of equations.

Check

Solve the system of equations by graphing.

$y + \dfrac{1}{3}x = 1$

$y = -\dfrac{1}{3}x + 1$

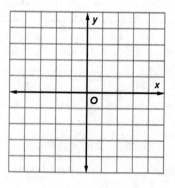

Go Online You can complete an Extra Example online.

🌍 Apply Bake Sale

The Spanish Club is having a bake sale. You can buy a bag of trail mix and a loaf of pumpkin bread for $11 or three bags of trail mix and two loaves of pumpkin bread for $24. This situation can be represented with the system $x + y = 11$ and $3x + 2y = 24$, where x represents the cost of a bag of trail mix and y represents the cost of a loaf of pumpkin bread. How much does each bag of trail mix and each loaf of pumpkin bread cost?

1 What is the task?

Make sure you understand exactly what question to answer or problem to solve. You may want to read the problem three times. Discuss these questions with a partner.

First Time Describe the context of the problem, in your own words.
Second Time What mathematics do you see in the problem?
Third Time What are you wondering about?

2 How can you approach the task? What strategies can you use?

Record your observations here

3 What is your solution?

Use your strategy to solve the problem.

Show your work here

💬 Talk About It!

How do you know the solution cannot be $9 for each bag of trail mix and $2 for each loaf of pumpkin bread?

4 How can you show your solution is reasonable?

✍ **Write About It!** Write an argument that can be used to defend your solution.

Check

Chloe skated an average speed of 10 miles per hour, while Avery skated an average speed of 5 miles per hour. They skated a total of 20 miles in 2.5 hours. This situation can be represented with the system $x + y = 2.5$ and $10x + 5y = 20$, where x represents the number of hours Chloe skated and y represents the number of hours Avery skated. How long did each person skate?

Show your work here

🐦 **Go Online** You can complete an Extra Example online.

📖 **Foldables** It's time to update your Foldable, located in the Module Review, based on what you learned in this lesson. If you haven't already assembled your Foldable, you can find the instructions on page FL1.

Solve Systems of Equations

one solution	no solution	infinite number of solutions

Solve Algebraically	Solve Algebraically	Solve Algebraically
Example	Example	Example

Practice

⟡ **Go Online** You can complete your homework online.

Solve each system of equations by graphing. Check the solution. (Examples 1–4)

1. $y = x + 4$
$y = -2x - 2$

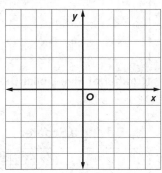

2. $y - \frac{1}{2}x = -1$
$y = \frac{1}{2}x + 4$

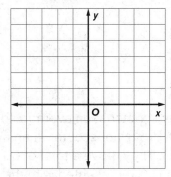

3. $y + \frac{1}{4}x = 1$
$y = -\frac{1}{4}x + 1$

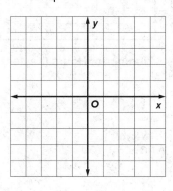

4. $x = -3$
$y = 5$

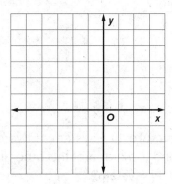

Test Practice

5. Grid The graph of a system of equations is shown. Plot and label the solution of the system on the graph.

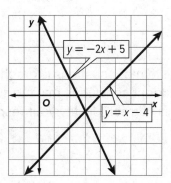

$y = -2x + 5$

$y = x - 4$

Apply

6. Salim has 9 five-dollar bills and ten-dollar bills in his wallet. The total amount in his wallet is $60. This situation can be represented with the system $x + y = 9$ and $5x + 10y = 60$, where x represents the number of five-dollar bills and y represents the number of ten-dollar bills. How many five- and ten-dollar bills does Salim have?

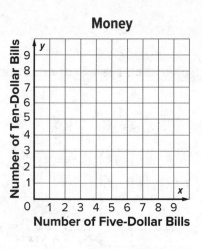

Money

7. Kaylee drove her scooter at an average speed of 15 miles per hour. Sophia drove her scooter at an average speed of 10 miles per hour. They drove a total of 60 miles in 5 hours. This situation can be represented with the system $x + y = 5$ and $15x + 10y = 60$, where x represents the number of hours Kaylee drove her scooter and y represents the number of hours Sophia drove her scooter. How long did each person drive their scooter?

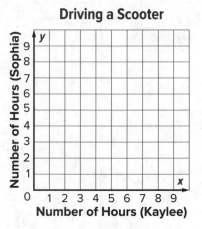

Driving a Scooter

8. 🔵 **Identify Structure** In a system of two equations, the graph of one line has a positive slope and the graph of the other line has a negative slope. What can you conclude about the system?

9. The system of equations $y = 2(x + 3)$ and $y - 2x = 6$ has infinitely many solutions. Explain why the point $(-1, -5)$ is not a solution of the system.

10. A system of equations is graphed on the coordinate plane. A student concludes that the solution of the system is $(-0.5, 1.5)$. Is this correct? Justify your response.

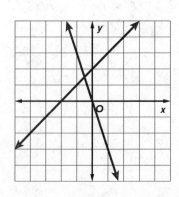

Determine Number of Solutions

I Can... use the slope-intercept form of lines in order to determine whether a system of equations has zero, one, or infinitely many solutions.

Explore Systems of Equations: Slopes and *y*-Intercepts

Online Activity You will use Web Sketchpad to explore the relationship between the slopes and *y*-intercepts of systems of equations and the number of solutions to the system.

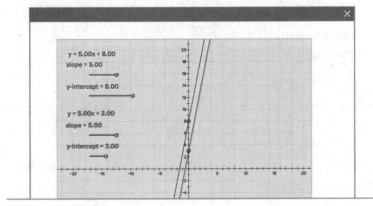

Pause and Reflect

Compare and contrast the graph of a system of equations with one solution to the graph of a system of equations with no solution. Include discussion about slopes and *y*-intercepts in your answer.

Record your observations here

Learn Systems of Equations: Compare Slopes and *y*-Intercepts

You can determine the number of solutions of a system of equations by comparing the slopes and *y*-intercepts.

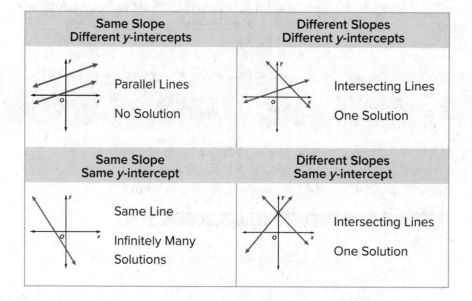

Same Slope Different *y*-intercepts	Different Slopes Different *y*-intercepts
Parallel Lines No Solution	Intersecting Lines One Solution

Same Slope Same *y*-intercept	Different Slopes Same *y*-intercept
Same Line Infinitely Many Solutions	Intersecting Lines One Solution

🫐 Think About It!

What do you know about the slope and *y*-intercepts when a system has no solution? one solution? an infinite number of solutions?

Example 1 Find the Number of Solutions

Determine if the system of equations has *no solution, one solution*, or *an infinite number of solutions*.

$y = -2x + 4$

$y = \frac{3}{4}x + 2$

Are the slopes the same?

Are the *y*-intercepts the same?

Since the lines have different slopes and different *y*-intercepts, they intersect in exactly _____ point. So, the system of equations has one solution.

💬 Talk About It!

Check the solution by graphing the system. How does the graph illustrate the solution?

Check

Does the system of equations have *no solution, one solution*, or *an infinite number of solutions*?

$y = 5 - x$

$y = 3x$

Show your work here

🅑 **Go Online** You can complete an Extra Example online.

Example 2 Find the Number of Solutions

Determine if the system of equations has *no solution, one solution,* or *an infinite number of solutions.*

$$y = \frac{2}{3}x + 3$$

$$3y = 2x + 15$$

Step 1 Write both equations in slope-intercept form.

The equation $y = \frac{2}{3}x + 3$ is already in slope-intercept form. You need to write $3y = 2x + 15$ in slope-intercept form.

$3y = 2x + 15$ Write the equation.

$$\frac{3y}{\boxed{}} = \frac{2x + 15}{\boxed{}}$$ Divide each side by 3.

$y = \boxed{} + \boxed{}$ Simplify.

Step 2 Analyze the equations.

$$y = \frac{2}{3}x + 3$$

$$y = \frac{2}{3}x + 5$$

The equations have _____ slopes. The equations have

_____ *y*-intercepts. So, the lines are _____ and

there is no solution of this system.

Check

Does the system of equations have *no solution, one solution,* or *an infinite number of solutions?*

$$y = -\frac{2}{5}x$$

$$y - 9 = -\frac{2}{5}x$$

Show your work here

 Go Online You can complete an Extra Example online.

Think About It!

Are both equations written in a form in which it is easy to compare the slopes and *y*-intercepts?

Talk About It!

Check the solution by graphing the system. How does the graph illustrate the solution?

Example 3 Find the Number of Solutions

Think About It!

How would you begin solving the problem?

Determine if the system of equations has _no solution, one solution,_ or _an infinite number of solutions._

$$y = \frac{3}{4}x$$
$$3x - 4y = 0$$

Step 1 Write both equations in slope-intercept form.

The equation $y = \frac{3}{4}x$ is already in slope-intercept form. Write the equation $3x - 4y = 0$ in slope-intercept form.

$3x - 4y = 0$		Write the equation.
$\underline{-3x \qquad = -3x}$		Subtract $3x$ from each side.
$\boxed{} = \boxed{}$		Simplify.
$\dfrac{-4y}{\boxed{}} = \dfrac{-3x}{\boxed{}}$		Divide each side by -4.
$y = \boxed{}$		Simplify.

Step 2 Analyze the equations.

$$y = \frac{3}{4}x$$
$$y = \frac{3}{4}x$$

The equations have _____ slopes. The equations have _____ y-intercepts. So, the lines are the same line and there are infinitely many solutions of this system.

Talk About It!

Describe a method you can use to verify the system has infinitely many solutions.

Check

Does the system of equations have _no solution, one solution,_ or _an infinite number of solutions?_

$$y - 8 = 2x$$
$$2y = 16 + 4x$$

Show your work here

Go Online You can complete an Extra Example online.

Pause and Reflect

Review Examples 1–3. What are the steps you would take to determine how many solutions a system of equations has?

Record your observations here

Example 4 Analyze Systems of Equations

A system of equations consists of two lines. One line passes through (0, 4) and (3, 2). The other line passes through (3, −3) and (9, −7).

Determine whether the line through the first pair of points intersects the line through the second pair of points.

Think About It!

How would you begin solving the problem?

Step 1 Find the slope of each line.

Find the slope of the line that passes through (0, 4) and (3, 2).

$$m = \frac{y_2 - y_1}{x_2 - x_1}$$ Slope formula

$$m = \frac{\boxed{}}{\boxed{}}$$ $(x_1, y_1) = (0, 4); (x_2, y_2) = (3, 2)$

$$m = \boxed{}$$ Simplify.

Find the slope of the line that passes through (3, −3) and (9, −7).

$$m = \frac{y_2 - y_1}{x_2 - x_1}$$ Slope formula

$$m = \frac{\boxed{}}{\boxed{}}$$ $(x_1, y_1) = (3, -3); (x_2, y_2) = (9, -7)$

$$m = -\frac{4}{6} \text{ or } \boxed{}$$ Simplify.

The lines have the same slopes.

(continued on next page)

Step 2 Find the y-intercept for each line.

The y-intercept is the y value when $x = 0$. So, the y-intercept of the line that passes through (0, 4) and (3, 2) is 4.

Use the slope-intercept form of a linear equation to find the y-intercept b of the line that passes through (3, −3) and (9, −7).

$y = mx + b$ Slope-intercept form

$y = \boxed{} x + b$ Replace m with $-\dfrac{2}{3}$.

$\boxed{} = -\dfrac{2}{3}\left(\boxed{}\right) + b$ Use the point (3, −3). $x = 3, y = -3$

$-3 = \boxed{} + b$ Simplify.

$\boxed{} = b$ Add 2 to each side.

The lines have different y-intercepts.

Step 3 Analyze the slopes and y-intercepts.

Since the slopes are the same and the y-intercepts are different, the lines are _____. So, the line through the first pair of points does *not* intersect the line through the second pair of points.

Check

A system of equations consists of two lines. One line passes through (0, 6) and (2, 2). The other line passes through (−1, −8) and (3, 4).

Does the line through the first pair of points intersect the line through the second pair of points?

 Apply Hiking

Two friends are hiking along linear paths at a state park. On a map of the trails, Grayson's path passes through the coordinates (2, 3) and (0, 5). Xander's path passes through the coordinates (1, 1) and (0, −1). Do the friends pass through a common coordinate, hike along the same path, or never cross paths?

1 What is the task?

Make sure you understand exactly what question to answer or problem to solve. You may want to read the problem three times. Discuss these questions with a partner.

First Time Describe the context of the problem, in your own words.
Second Time What mathematics do you see in the problem?
Third Time What are you wondering about?

2 How can you approach the task? What strategies can you use?

 Record your observations here

3 What is your solution?

Use your strategy to solve the problem.

 Show your work here

4 How can you show your solution is reasonable?

✏️ **Write About It!** Write an argument that can be used to defend your solution.

💬 Talk About It!

How can you solve the problem another way?

Check

Ellie and Samuel are riding bikes along linear routes in their town. On a map of their town, Ellie's route passes through the coordinates (−1, 3) and (0, 1). Samuel's route passes through the coordinates (1, 4) and (0, 2). Do the friends pass through a common coordinate, ride along the same route, or never cross routes?

Show your work here

R Go Online You can complete an Extra Example online.

Pause and Reflect

How well do you understand the concepts from today's lesson? What questions do you still have? How can you get those questions answered?

Record your observations here

Practice

Go Online You can complete your homework online.

Determine if each system of equations has *no solution, one solution*, or *an infinite number of solutions.* (Examples 1–3)

1. $-5x + y = -1$
$-5x + y = 10$

2. $y = -4x + 9$
$y = \frac{2}{3}x - 5$

3. $y + 1 = 3x$
$2y = 6x - 2$

4. $y = -\frac{4}{5}x$
$4x + 5y = 0$

5. $y = \frac{1}{2}x + 6$
$2y = x - 8$

6. $y = -2x$
$y = x + 3$

A system of equations consists of two lines. A line passes through each pair of points. Determine whether the line through the first pair of points intersects the line through the second pair of points. (Example 4)

7. (0, −5) and (2, −4);
(4, −3) and (6, −2)

8. (0, 4) and (1, 7);
(−1, −5) and (5, 13)

9. (0, 2) and (9, −1);
(12, 7) and (−6, −5)

Test Practice

10. Equation Editor Enter values for *a* and *b*, so that the system of equations has one solution.
$y = -6x - 4$
$y = ax + b$

$a = $ [_____]

$b = $ [_____]

Apply

11. Ethan and Camila are walking along linear routes in their town. On a map of their town, Ethan's route passes through the coordinates (0, −2) and (5, 18). Camila's route passes through the coordinates (−3, −12) and (0, 3). Do the friends pass through a common coordinate, walk along the same route, or never cross routes?

12. The Taylor family and the King family are each camping at a different national park. Let x represent the number of nights camping and y represent the total cost ($). The linear equation that represents the Taylor's total cost passes through the points (0, 15) and (5, 90). The linear equation that represents the King's total cost passes through the points (0, 0) and (7, 105). Determine if the national parks ever charge the same amount for a certain number of nights, always charge the same amount, or never charge the same amount for the same number of nights.

13. (MP) **Identify Structure** Without comparing the slopes and y-intercepts of the system shown, explain why the system of equations has no solution.

$$-7x + 3y = -5$$
$$-7x + 3y = 9$$

14. Which One Doesn't Belong? Circle the system of equations that does not belong with the other two. Explain your reasoning.

$y = x + 6$	$3x + y = -1$	$y = -4x - 3$
$y = -x + 2$	$y = 4x + 6$	$y + 4x = -5$

15. (MP) **Identify Structure** Systems of equations that have the same slope and different y-intercepts are parallel lines. How do the slopes of a system of equations that consist of perpendicular lines (lines that intersect at right angles) compare?

16. Write a system of equations that has one solution. Then describe how you can modify one of the equations in your system so that it has infinitely many solutions.

Solve Systems of Equations by Substitution

I Can... use substitution to solve a system of linear equations, including those that have zero or infinitely many solutions.

Explore Solve Systems of Equations by Substitution

Online Activity You will explore how to solve systems of equations by using the substitution method.

Complete the table to show the number of bags of pretzels, $3x$ and $20 - x$, for each granola bar purchased.

Talk About It!

How does the table illustrate the solution? How can you relate the expressions $3x$ and $20 - x$ to create an equation that, when solved, gives the solution?

Granola Bars, x	$3x$	$20 - x$
1	3	19
2		
3		
4		
5		
6		
7		
8		

Learn Solve Systems of Equations by Substitution

Substitution is an algebraic model that can be used to find the exact solution of a system of equations. It involves replacing one of the variables in one equation with an equivalent algebraic expression from the other equation, in order to solve for one of the variables.

Go Online Watch the animation to see how to solve the system $y = 4x - 7$ and $y = 2x + 5$ using substitution and these steps.

Step 1 Solve one equation for a variable.
In this example, both equations are already solved for y.

Step 2 Substitute the expression into the other equation.
$$y = 2x + 5$$
$$4x - 7 = 2x + 5 \qquad \text{Replace } y \text{ with } 4x - 7.$$

Step 3 Solve the equation to find the value of a variable.
$$4x - 7 = 2x + 5$$
$$\underline{-2x \qquad = -2x} \qquad \text{Subtract } 2x \text{ from each side.}$$
$$2x - 7 = 5$$
$$\underline{+7 = +7} \qquad \text{Add 7 to each side.}$$
$$\frac{2x}{2} = \frac{12}{2} \qquad \text{Divide each side by 2.}$$
$$x = 6$$

(continued on next page)

Step 4 Substitute the value of the variable into one of the original equations and solve.

$y = 4x - 7$ Write the equation.
$y = 4(6) - 7$ Replace x with 6.
$y = 17$ Simplify.

Step 5 Check that both equations are true when $x = 6$ and $y = 17$, and write the solution as an ordered pair.

$y = 4x - 7$	$y = 2x + 5$	Write the equation.
$(17) \stackrel{?}{=} 4(6) - 7$	$(17) \stackrel{?}{=} 2(6) + 5$	Replace x with 6 and y with 17.
$17 = 17$	$17 = 17$	Simplify. Both sentences are true.

Since both sentences are true, the solution of the system of equations is (6, 17).

Example 1 Solve Systems by Substitution

Solve the system of equations by substitution.
$y = x - 3$
$y = 2x$

Step 1 Since y is equal to $2x$, replace y with $2x$ in the other equation, $y = x - 3$. Then solve the equation.

$y = x - 3$ Write the equation.

$\boxed{} = x - 3$ Replace y with $2x$.

$\underline{-x = -x}$ Subtraction Property of Equality

$\boxed{} = \boxed{}$ Simplify.

Step 2 Since $x = -3$, substitute -3 for x in either equation to find the value of y.

$y = x - 3$ Write the first equation.

$y = \boxed{} - 3$ Replace x with -3.

$y = \boxed{}$ Simplify.

So, the solution of this system of equations is $(-3, -6)$.

Check

Solve the system of equations by substitution.
$y = -6 + x$
$y = 3x$

Show your work here

Learn Rewrite Equations to Solve Systems by Substitution

Sometimes one or both equations in a system may not be written in slope-intercept form. When solving a system by substitution, one of the equations may need to be solved for either x or y. This is important in order to be able to eliminate one of the variables.

The equation $2x + y = -10$ is solved for x and solved for y, shown below.

$$x = -5 - \frac{1}{2}y \qquad\qquad y = -10 - 2x$$

Example 2 Rewrite One Equation to Solve Systems by Substitution

Solve the system of equations by substitution.
$$y = 3x + 8$$
$$8x + 4y = 12$$

Step 1 Since y is equal to $3x + 8$, replace y with $3x + 8$ in the other equation, $8x + 4y = 12$. Then solve the equation.

$8x + 4y = 12$	Write the equation.
$8x + 4\left(\boxed{}\right) = 12$	Replace y with $3x + 8$.
$8x + \boxed{} + \boxed{} = 12$	Distributive Property
$\boxed{} + 32 = 12$	Combine like terms.
$20x = \boxed{}$	Subtract 32 from each side.
$x = \boxed{}$	Divide each side by 20.

Step 2 Since $x = -1$, substitute -1 for x in either equation to find the value of y.

$y = 3x + 8$	Write the equation.
$y = 3\left(\boxed{}\right) + 8$	Replace x with -1.
$y = \boxed{}$	Simplify.

So, the solution of this system of equations is $(-1, 5)$.

 Think About It!

Is one of the equations solved for either x or y? How will this help you eliminate one of the variables?

 Talk About It!

In Step 2, you substituted -1 for x into the equation $y = 3x + 8$. You can also substitute -1 for x into the other equation $8x + 4y = 12$. Why is either method correct? Which do you prefer? Explain.

Solve the system of equations by substitution.

$2x + 5y = 44$

$y = 6x - 4$

Show your work here

 Go Online You can complete an Extra Example online.

Example 3 Rewrite Both Equations to Solve Systems by Substitution

Solve the system of equations by substitution.

$y - 3x = -13$

$4x + 5y = 11$

Step 1 Solve the equation $y - 3x = -13$ for y.

$y - 3x = -13$	Write the equation.
$\underline{+\,3x = +\,3x}$	Addition Property of Equality
$y = \boxed{}$	Simplify.

Step 2 Since y is equal to $-13 + 3x$, replace y with $-13 + 3x$ in the other equation, $4x + 5y = 11$. Then solve the equation.

$4x + 5y = 11$	Write the equation.
$4x + 5\left(\boxed{}\right) = 11$	Replace y with $-13 + 3x$.
$4x - \boxed{} + \boxed{} = 11$	Distributive Property
$\boxed{} - 65 = 11$	Combine like terms.
$19x = \boxed{}$	Add 65 to each side.
$x = \boxed{}$	Divide each side by 19.

Step 3 Since $x = 4$, substitute 4 for x in either equation to find the value of y.

$y - 3x = -13$	Write the equation.
$y - 3\left(\boxed{}\right) = -13$	Replace x with 4.
$y - \boxed{} = -13$	Simplify.
$y = \boxed{}$	Add 12 to each side.

So, the solution of this system of equations is $(4, -1)$.

 Think About It!

Is one of the equations solved for either x or y? Which equation will be easier to solve, and for which variable?

 Talk About It

In Step 3, you substituted 4 for x into the equation $y - 3x = -13$. You can also substitute 4 for x into the other equation $4x + 5y = 11$. Why is either method correct? Compare the methods.

Check

Solve the system of equations by substitution.

$x + 4y = 3$

$2x - 3y = 17$

Show your work here

Example 4 Solve Systems with Infinitely Many Solutions

Solve the system of equations by substitution.

$y = x + 5$

$2y - 2x = 10$

Since y is equal to $x + 5$, replace y with $x + 5$ in the equation $2y - 2x = 10$. Then solve the equation.

$2y - 2x = 10$ Write the equation.

$2\left(\boxed{}\right) - 2x = 10$ Replace y with $x + 5$.

$\boxed{} + \boxed{} - 2x = 10$ Distributive Property

$\boxed{} = 10$ Simplify.

$10 = 10$ is a true statement. So, there is an infinite number of solutions.

Check

Solve the system of equations by substitution.

$\frac{1}{2}x - y = 2$

$x = 2y + 4$

Show your work here

Go Online You can complete an Extra Example online.

 Think About It!

Compare the like terms in the system of equations. What do you notice?

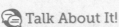 **Talk About It!**

Without using substitution, how can you rewrite the equation $2y - 2x = 10$ to verify that the system has infinitely many solutions?

Example 5 Solve Systems with No Solutions

Solve the system of equations by substitution.

$x - y = -4$

$x - y = \frac{2}{5}$

Step 1 Solve either equation for x.

$x - y = -4$ Write the equation.

$\underline{ + y = + y }$ Addition Property of Equality

$x = \boxed{}$ Simplify.

Step 2 Since x is equal to $-4 + y$, replace x with $-4 + y$ in the other equation, $x - y = \frac{2}{5}$. Then solve the equation.

$x - y = \frac{2}{5}$ Write the equation.

$\left(\boxed{}\right) - y = \frac{2}{5}$ Replace x with $-4 + y$.

$\boxed{} = \frac{2}{5}$ Simplify.

The statement $-4 = \frac{2}{5}$ is never true. So, there is no solution.

Check

Solve the system of equations by substitution.

$7x + y = 9$
$y + 7x = -2$

Show your work here

Go Online You can complete an Extra Example online.

🌐 Apply Greeting Cards

Mei paid $15.75 for 6 greeting cards. Some of the cards cost
$2.50 each, and some cost $3.25 each. Let x represent the number
of cards that cost $2.50 each, and y represent the number of cards
that cost $3.25 each. This situation can be represented by the
system $x + y = 6$ and $2.5x + 3.25y = 15.75$. How many of each type
of card did Mei purchase?

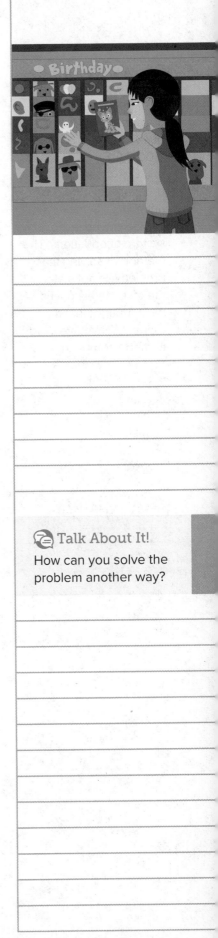

1 What is the task?

Make sure you understand exactly what question to answer or
problem to solve. You may want to read the problem three times.
Discuss these questions with a partner.

First Time Describe the context of the problem, in your own words.
Second Time What mathematics do you see in the problem?
Third Time What are you wondering about?

2 How can you approach the task? What strategies can you use?

3 What is your solution?

Use your strategy to solve the problem.

🗨 **Talk About It!**
How can you solve the
problem another way?

4 How can you show your solution is reasonable?

✏️ **Write About It!** Write an argument that can be used to defend
your solution.

Check

The length of a rectangle is 3 meters more than the width. The perimeter is 26 meters. This situation can be represented with the system $\ell - w = 3$ and $2\ell + 2w = 26$, where ℓ represents the length of the rectangle and w represents the width. Find the dimensions of the rectangle.

Show your work here

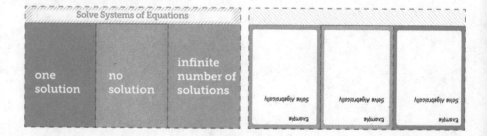

Go Online You can complete an Extra Example online.

Foldables It's time to update your Foldable, located in the Module Review, based on what you learned in this lesson. If you haven't already assembled your Foldable, you can find the instructions on page FL1.

Solve Systems of Equations

one solution	no solution	infinite number of solutions

Practice

Solve each system of equations by substitution. Check the solution.
(Examples 1–5)

1. $y = x - 14$
$\quad y = -6x$

2. $x - y = -5$
$\quad x - y = \frac{1}{3}$

3. $y + 7 = 2x$
$\quad 2y = 4x - 14$

4. $y - 6x = 12$
$\quad y = 6x + 5$

5. $y = 3x - 7$
$\quad 4x + y = -14$

6. $y = -6x + 8$
$\quad 2y + 12x = 16$

7. $-3x + 4y = 6$
$\quad -x + 2y = 8$

8. $y + 11 = 2x$
$\quad 3y - 6x = -33$

9. $9x + y = 9$
$\quad y + 9x = 5$

Test Practice

10. Solve the system of equations by substitution.

$y = \frac{1}{4}x - 1$

$2y = \frac{2}{3}x + 6$

11. Open Response What is the solution of the system of equations?

$y = 2x - 4$
$-21x + 3y = 3$

Apply

12. In one basketball game, Zaid made 7 two-point and three-point baskets to score 17 points. Let x represent the number of two-point baskets, and y represent the number of three-point baskets. This situation can be represented by the system $x + y = 7$ and $2x + 3y = 17$. How many of each type of basket did Zaid make?

13. The length of a rectangle is 4.6 feet more than the width. The perimeter is 38 feet. This situation can be represented with the system $\ell - w = 4.6$ and $2\ell + 2w = 38$, where ℓ represents the length of the rectangle and w represents the width. Find the dimensions of the rectangle.

14. MP Be Precise Describe the difference between solving a system of equations by graphing and solving by substitution.

15. MP Find the Error The first steps for solving the system of equations $y = x - 1$ and $y - x = -1$ by substitution are shown. A student concludes that there is no solution. Find the mistake and correct it.

$$y - x = -1$$
$$(x - 1) - x = -1$$
$$-1 = -1$$

16. Order the steps to solve a system of equations by substitution.

_____ Write the x- and y-values as an ordered pair.

_____ Solve the equation to find the value of a variable.

_____ Solve one equation for a variable.

_____ Substitute the value of the variable into one of the original equations and solve.

_____ Substitute the expression into the other equation.

17. When solving a system of equations by substitution, does one of the two equations *always*, *sometimes*, or *never* need to be written in slope-intercept form? Explain your reasoning.

Solve Systems of Equations by Elimination

I Can... use elimination to solve a system of linear equations.

What Vocabulary Will You Learn?
elimination

Explore Solve Systems of Equations by Elimination

Online Activity You will explore how to solve systems of equations by eliminating one of the variables.

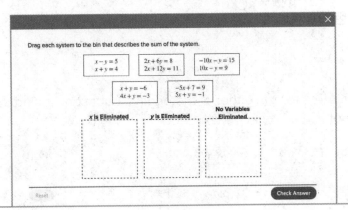

Learn Solve Systems by Elimination: Addition

The use of addition or subtraction to eliminate one variable in a system of equations is called **elimination**. Elimination is an algebraic method you can use to solve a system of equations.

Go Online Watch the animation to see how to solve the system $x + 3y = -3$ and $4x - 3y = 18$ using elimination and these steps.

Because the coefficients of y, 3 and -3 are opposites, add the equations to eliminate the variable.

Steps 1 and 2 Add the equations. Then solve for one of the variables.

$$x + 3y = -3$$
$$\underline{(+)\ 4x - 3y = 18}$$
$$5x + 0 = 15 \quad \text{Add; the variable } y \text{ is eliminated.}$$
$$x = 3 \quad \text{Divide each side by 5.}$$

Step 3 Substitute the value of the variable into one of the original equations and solve for the other variable.

$$x + 3y = -3 \quad \text{Write the equation.}$$
$$3 + 3y = -3 \quad \text{Replace } x \text{ with 3.}$$
$$y = -2 \quad \text{Solve the equation.}$$

Step 4 Write the x- and y-values as an ordered pair. The solution of the system of equations is $(3, -2)$.

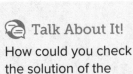

Talk About It!

How could you check the solution of the system by graphing?

Think About It!

Which terms, if any, have opposite coefficients? How can this help you decide how to eliminate one of the variables?

Talk About It!

In Step 2, you substituted 4 for x into the equation $2x + 5y = 3$. You can also substitute 4 for x into the other equation $4x - 5y = 21$. Explain why either method is correct.

Example 1 Solve Systems by Elimination: Addition

Solve the system of equations by elimination.

$$2x + 5y = 3$$
$$4x - 5y = 21$$

Step 1 Eliminate the variable y by adding the equations.

$$2x + 5y = 3 \qquad \text{Align like terms.}$$
$$\underline{(+)\ 4x - 5y = 21}$$
$$\boxed{} = \boxed{} \qquad \text{Add; the variable } y \text{ is eliminated.}$$
$$\frac{6x}{6} = \frac{24}{6} \qquad \text{Division Property of Equality}$$
$$x = \boxed{} \qquad \text{Simplify.}$$

Step 2 Substitute 4 for x in either of the original equations to find the value of y.

$$2x + 5y = 3 \qquad \text{Write the equation.}$$
$$2\left(\boxed{}\right) + 5y = 3 \qquad \text{Replace } x \text{ with 4.}$$
$$\boxed{} + 5y = 3 \qquad \text{Simplify.}$$
$$5y = \boxed{} \qquad \text{Subtract 8 from each side.}$$
$$y = \boxed{} \qquad \text{Divide each side by 5.}$$

So, the solution of this system is (4, −1).

Check

Solve the system of equations by elimination.

$$4x - y = -8$$
$$7x + y = -14$$

Show your work here

Go Online You can complete an Extra Example online.

Learn Solve Systems by Elimination: Subtraction

When the coefficients of a variable are the same, *subtracting* the equations will eliminate the variable. When subtracting integers, you add the opposite. The same process is used when subtracting one equation from another.

⬣ **Go Online** Watch the animation to see how to solve the system $4x + 5y = 18$ and $-2x + 5y = 6$ by elimination using subtraction and these steps.

Step 1 Subtract the equations.

Subtraction is the opposite of addition, so add the opposite of $-2x + 5y = 6$.

$$
\begin{array}{rcl}
4x + 5y = 18 \\
-(-2x + 5y = 6)
\end{array}
\rightarrow
\begin{array}{rcl}
4x + 5y = 18 \\
(+) \quad 2x - 5y = -6 \\
\hline
\boxed{} = \boxed{}
\end{array}
$$

Step 2 Solve for one of the variables using the resulting equation.

$6x = 12$	Write the equation.
$\dfrac{6x}{6} = \dfrac{12}{6}$	Divide each side by 6.
$x = \boxed{}$	Simplify.

Step 3 Substitute the value of the variable into one of the original equations and solve.

$-2x + 5y = 6$	Write the equation.
$-2\left(\boxed{}\right) + 5y = 6$	Replace x with 2.
$\boxed{} + 5y = 6$	Simplify.
$\underline{+4 \qquad = +4}$	Add 4 to each side.
$\dfrac{5y}{5} = \dfrac{10}{5}$	Divide each side by 5.
$y = \boxed{}$	Simplify.

Step 4 Write the x- and y-values as an ordered pair.

The solution of the system of equations is (2, 2).

💬 **Talk About It!**

In Step 2, a student mistakenly wrote $x = 12$. What was the student's solution of the system? How could the student recognize that this solution is incorrect?

Example 2 Solve Systems by Elimination: Subtraction

Solve the system of equations by elimination.

$8x + 2y = 12$
$8x + 4y = 16$

Step 1 Subtract the equations to eliminate the variable x.

Subtraction is the opposite of addition, so add the opposite of $8x + 4y = 16$. The opposite of $8x + 4y = 16$ is $-8x - 4y = -16$.

$$8x + 2y = 12$$
$$\underline{(+) -8x - 4y = -16}$$ Align like terms.

$$\boxed{} = \boxed{}$$ The variable x is eliminated.

$$\frac{-2y}{-2} = \frac{-4}{-2}$$ Division Property of Equality

$$y = \boxed{}$$ Simplify.

Step 2 Substitute 2 for y in either of the original equations to find the value of x.

$$8x + 2y = 12$$ Write the equation.

$$8x + 2\left(\boxed{}\right) = 12$$ Replace y with 2.

$$8x + \boxed{} = 12$$ Simplify.

$$8x = \boxed{}$$ Subtract 4 from each side.

$$x = \boxed{}$$ Divide each side by 8.

So, the solution of this system of equations is (1, 2).

Check

Solve the system of equations by elimination.

$-6x + 5y = 26$
$-6x + 2y = 14$

Show your work here

 Go Online You can complete an Extra Example online.

Think About It!
If you add the two equations, will any variables be eliminated? Explain.

Talk About It!
In Step 2, you substituted 2 for y into the equation $8x + 2y = 12$. You can also substitute 2 for y into the other equation $8x + 4y = 16$. Why is either method correct? Compare the methods.

Learn Solve Systems by Elimination: Multiplication

When neither variable in a system of equations can be eliminated by adding or subtracting, you can use multiplication.

Go Online Watch the animation to see how to solve the system $3x + 2y = 12$ and $-x + 4y = 10$ by elimination using multiplication. The animation shows these steps.

Step 1 Multiply each term in one equation by a constant to create opposite coefficients.

$$3x + 2y = 12 \quad \rightarrow \quad 3x + 2y = 12$$
$$3(-x + 4y = 10) \quad \rightarrow \quad \boxed{} = \boxed{}$$

The equations now have opposite coefficients of x, 3 and -3.

Step 2 Add the equations and solve for one of the variables.

$$
\begin{array}{rcl}
3x + 2y &=& 12 \\
(+) -3x + 12y &=& 30 \\
\hline
\boxed{} &=& \boxed{}
\end{array}
$$
    Add; the variable x is eliminated.

$$\frac{14y}{14} = \frac{42}{14}$$
    Divide each side by 14.

$$y = \boxed{}$$
    Simplify.

Step 3 Substitute the value of the variable into one of the original equations and solve.

$$3x + 2y = 12$$
    Write the equation.

$$3x + 2\left(\boxed{}\right) = 12$$
    Replace y with 3.

$$3x + \boxed{} = 12$$
    Simplify.

$$
\begin{array}{rcl}
-6 &=& -6
\end{array}
$$
    Subtract 6 from each side.

$$\frac{3x}{3} = \frac{6}{3}$$
    Divide each side by 3.

$$x = \boxed{}$$
    Simplify.

Step 4 Write the x- and y-values as an ordered pair.

The solution of the system of equations is (2, 3).

Talk About It!

What property of equality allows you to multiply each side of an equation by the same constant?

Example 3 Multiply One Equation to Eliminate a Variable

Think About It!
If you add or subtract the two equations, will any variables be eliminated? Explain.

Solve the system of equations by elimination. Check the solution.

$2x + 5y = -13$
$x + 3y = -5$

Step 1 Multiply one equation by a constant.

What constant can you multiply the equation $x + 3y = -5$ by to create opposite x coefficients? _____

$$2x + 5y = -13 \quad\quad \rightarrow \quad\quad 2x + 5y = -13$$

$$-2(x + 3y) = -2(-5) \quad \rightarrow \quad \boxed{} = \boxed{}$$

Step 2 Add the equations to eliminate a variable. Then solve the equation.

$$2x + 5y = -13$$
$$\underline{(+) -2x - 6y = 10} \quad\quad \text{Align like terms.}$$
$$\boxed{} = \boxed{} \quad\quad \text{Add; the variable } x \text{ is eliminated.}$$
$$\frac{-1y}{-1} = \frac{-3}{-1} \quad\quad \text{Division Property of Equality}$$
$$y = \boxed{} \quad\quad \text{Simplify.}$$

Step 3 Substitute 3 for y in either of the original equations to find the value of x.

$$x + 3y = -5 \quad\quad \text{Write the equation.}$$
$$x + 3\left(\boxed{} \right) = -5 \quad\quad \text{Replace } y \text{ with 3.}$$
$$x + \boxed{} = -5 \quad\quad \text{Simplify.}$$
$$x = \boxed{} \quad\quad \text{Subtract 9 from each side.}$$

So, the solution of this system of equations is $(-14, 3)$.

(continued on next page)

Step 4 Check the solution.

Replace x with -14 and y with 3 in both of the original equations.

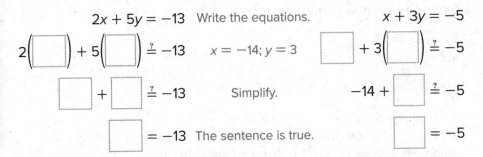

$2x + 5y = -13$ Write the equations.

$2\left(\boxed{}\right) + 5\left(\boxed{}\right) \overset{?}{=} -13$ $x = -14; y = 3$

$\boxed{} + \boxed{} \overset{?}{=} -13$ Simplify.

$\boxed{} = -13$ The sentence is true.

$x + 3y = -5$

$\boxed{} + 3\left(\boxed{}\right) \overset{?}{=} -5$

$-14 + \boxed{} \overset{?}{=} -5$

$\boxed{} = -5$

Since both of the sentences are true, the solution of the system of equations is $(-14, 3)$.

Check

Solve the system of equations by elimination.

$x + 2y = -1$
$7x - 3y = 10$

Show your work here

🡢 **Go Online** You can complete an Extra Example online.

Pause and Reflect

Review the three methods for solving systems of equations that you learned in this module: graphing, substitution, and elimination. Explain when it would be most advantageous to use each method.

Record your observations here

Talk About It!

In Step 1, you multiplied the equation $x + 3y = -5$ by -2, so that you can add the equations to eliminate the variable x. If you wanted to eliminate the variable y instead, what can you do? Which method is more efficient, in this case? Explain why.

Example 4 Multiply Both Equations to Eliminate a Variable

Solve the system of equations by elimination.

$6x + 2y = 14$
$4x + 3y = 1$

Step 1 Multiply both equations by constants.

Sometimes, you need to multiply both equations by constants in order to eliminate one of the variables. Multiply $6x + 2y = 14$ by 3 and $4x + 3y = 1$ by -2 in order to eliminate the variable y.

$$3(6x + 2y) = 3(14) \quad \rightarrow \quad 18x + 6y = 42$$
$$-2(4x + 3y) = -2(1) \quad \rightarrow \quad -8x + (-6y) = -2$$

Step 2 Add the equations to eliminate a variable. Then solve the equation.

$$
\begin{array}{ll}
18x + 6y = 42 & \\
\underline{(+) -8x - 6y = -2} & \text{Align like terms.} \\
10x + 0 = 40 & \text{Add; the variable } y \text{ is eliminated.} \\
x = \boxed{} & \text{Divide each side by 10.}
\end{array}
$$

Step 3 Substitute 4 for x in either of the original equations to find the value of y.

$$
\begin{array}{ll}
6x + 2y = 14 & \text{Write the equation.} \\
6(4) + 2y = 14 & \text{Replace } x \text{ with 4.} \\
24 + 2y = 14 & \text{Simplify.} \\
y = \boxed{} & \text{Solve the equation.}
\end{array}
$$

So, the solution of this system of equations is $(4, -5)$.

Check

Solve the system of equations by elimination.

$-6x - 4y = 6$
$5x + 3y = -4$

Go Online You can complete an Extra Example online.

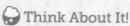 **Think About It!**

Will multiplying only one of the equations by a constant eliminate one of the variables? Explain.

Talk About It!

In this Example, the variable y was eliminated. How can the variable x be eliminated instead?

🌐 **Apply** Packaging

Andre is mailing packages that contain video games and DVDs. Three video games and one DVD weigh a total of 19 ounces. Five video games and two DVDs weigh a total of 33 ounces. This situation can be represented with the system $3x + y = 19$ and $5x + 2y = 33$, where x represents the weight of each video game and y represents the weight of each DVD. What is the weight of one video game and one DVD?

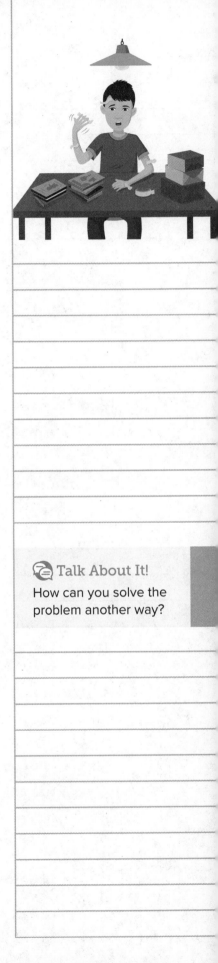

1 What is the task?

Make sure you understand exactly what question to answer or problem to solve. You may want to read the problem three times. Discuss these questions with a partner.

First Time Describe the context of the problem, in your own words.
Second Time What mathematics do you see in the problem?
Third Time What are you wondering about?

2 How can you approach the task? What strategies can you use?

3 What is your solution?

Use your strategy to solve the problem.

💬 Talk About It!
How can you solve the problem another way?

4 How can you show your solution is reasonable?

✏️ **Write About It!** Write an argument that can be used to defend your solution.

Check

The cost of 8 muffins and 2 quarts of milk is $18. The cost of 3 muffins and 1 quart of milk is $7.50. This situation can be represented with the system $8x + 2y = 18$ and $3x + y = 7.50$, where x represents the cost of each muffin and y represents the cost of one quart of milk. Find the cost of a muffin and a quart of milk.

Show your work here

Go Online You can complete an Extra Example online.

Pause and Reflect

What constitutes a solution of a system of equations?
Use mathematical language from this module in your explanation.

Record your observations here

Practice

Go Online You can complete your homework online.

Solve each system of equations by elimination. Check the solution.
(Examples 1–4)

1. $-6x + y = -3$
$5x - 2y = -8$

2. $-3x + 12y = 18$
$-6x + 24y = 36$

3. $-5x - 2y = -12$
$3x + 2y = 8$

4. $5x + 5y = -10$
$2x - 3y = -9$

5. $x + 3y = 6$
$x - 3y = 12$

6. $6x + 4y = 6$
$6x + 2y = 12$

7. $3x - 5y = 11$
$x - 4y = -8$

8. $-18x + 6y = -6$
$-24x + 6y = -18$

9. $-4x - 8y = 8$
$3x - 5y = 16$

Test Practice

10. Solve the system of equations by elimination.

$y = -\frac{1}{3}x - 5$

$\frac{1}{3}x + 5y = -9$

11. Open Response What is the solution of the system of equations?

$-2x + 4y = -10$
$2x + 2y = -8$

Apply

12. Jose and Caroline are shopping for school supplies. Caroline spent $15.50 on 5 notebooks and 3 folders. Jose bought 4 notebooks and 4 folders and spent $14. This situation can be represented with the system $5x + 3y = 15.50$ and $4x + 4y = 14$, where x represents the cost of each notebook and y represents the cost of each folder. What is the cost of one notebook and one folder?

13. The admission fee at a carnival is $5 for children and $8 for adults. On Friday, 1,250 people attended the carnival and $7,300 was collected. This situation can be represented with the system $5x + 8y = 7,300$ and $x + y = 1,250$, where x represents the number of children that attended the carnival and y represents the number of adults that attended the carnival. Find the number of children and the number of adults that attended the carnival on Friday.

14. **MP Identify Structure** Describe the structure of a system of equations for which it is more efficient to solve using elimination rather than substitution.

15. **MP Find the Error** The first steps for solving the system of equations by elimination using subtraction are shown. Find the mistake and correct it.

$$8x - y = -10 \quad \rightarrow \quad 8x - y = -10$$
$$6x - y = 4 \quad \rightarrow \quad \underline{-6x + y = 4}$$
$$2x + 0 = -6$$

16. Write a system of two equations that can be solved by elimination, in which one or both of the equations first needs to be multiplied by a constant. Explain how one of the variables can be eliminated.

17. **MP Identify Repeated Reasoning** Suppose the x-variables of a system of equations can be eliminated by adding the equations. What is the relationship between the coefficients? What is the value of a?

$$ax + 7y = c$$
$$5x + by = 12$$

Write and Solve Systems of Equations

I Can... write and solve a system of equations that models a real-world scenario.

Learn Write and Solve Systems of Equations

You can write a system of equations to solve real-world problems.

Benjamin and Harper went shopping at the school store. Benjamin spent $7 on 2 packages of pencils and 5 folders. Harper bought 1 package of pencils and 4 folders, and spent $4.25. What is the cost of a package of pencils and a folder?

Words
Describe the situation. Use only the most important words.
2 packages of pencils and 5 folders cost $7.
1 package of pencils and 4 folders cost $4.25.
Variables
Define variables to represent the unknown quantities.
Let x represent the cost of a package of pencils.
Let y represent the cost of a folder.
Equations
Translate your verbal model into algebraic equations.
Benjamin: $2x + 5y = 7$
Harper: $x + 4y = 4.25$

You can use graphing, substitution, or elimination to solve a system of equations. After representing the situation with a system, analyze the equations to determine which method is most efficient.

Pause and Reflect

Are there methods for solving systems of equations that you are not yet comfortable with using? What could you do to better understand each method?

Record your observations here

Talk About It!

Compare and contrast the methods you could use to solve the system $2x + 5y = 7$ and $x + 4y = 4.25$.

Example 1 Write and Solve Systems of Equations

The sum of two numbers is 39. Their difference is 7.

Write and solve a system of equations to find the two numbers.

Part A Write a system of equations.

Words
The sum of two numbers is _____.
The difference between the two numbers is _____.

Variables
Let x represent one number.
Let y represent the other number.

Equations
Sum: $\boxed{} + \boxed{} = \boxed{}$
Difference: $\boxed{} - \boxed{} = \boxed{}$

Part B Solve the system of equations.

Add the equations, eliminating the variable y. Then solve the equation.

$$
\begin{array}{ll}
x + y = 39 & \\
\underline{(+)\, x - y = 7} & \text{Align like terms.} \\
2x + 0 = 46 & \text{Add; the variable } y \text{ is eliminated.} \\
2x = 46 & \text{Simplify.} \\
x = \boxed{} & \text{Divide each side by 2.}
\end{array}
$$

Substitute 23 for x in either of the original equations to find the value of y.

$$y = \boxed{}$$

Part C Interpret the solution.

The solution is (23, 16). This means the numbers are 23 and 16.

💬 **Talk About It!**

Elimination was used to
solve this system of
equations. Explain why
this is the most efficient
method.

Check

The sum of two numbers is 14.5, and their difference is 5.5. Write and solve a system of equations to find the two numbers.

Part A Select the two equations that represent the situation, where x represents one number and y represents the other number.

(A) $14.5 + 5.5 = x$

(B) $x + y = 14.5$

(C) $x + y = 5.5$

(D) $x - y = 5.5$

(E) $x - y = 14.5$

Part B What is the solution of the system of equations?

Part C Interpret the solution.

Show your work here

🐦 **Go Online** You can complete an Extra Example online.

Pause and Reflect

Did you make any errors when completing the Check exercise? Without performing a check, how can you determine whether your solution is reasonable?

Record your observations here

Example 2 Write and Solve Systems of Equations

Creative Crafts offers scrapbooking lessons for $15 per hour plus a $10 supply charge. Scrapbooks Incorporated offers lessons for $20 per hour with no additional charges.

Write and solve a system of equations to determine for what number of hours the cost of lessons is the same at each store.

Part A Write a system of equations.

Words
The total cost at Creative Crafts is $_____ per hour plus a $_____ charge.
The total cost at Scrapbooks Incorporated is $_____ per hour.
Variables
Let x represent the number of hours.
Let y represent the total cost.
Equations
Creative Crafts: $y =$ ▭
Scrapbooks Inc.: $y =$ ▭

Part B Solve the system of equations.

Graph both equations on the same coordinate plane.

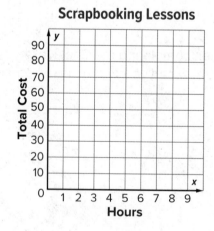

Scrapbooking Lessons

The graphs of the lines appear to intersect at (▭ , ▭).

(continued on next page)

Part C Verify and interpret the solution.

Check the solution by replacing x with 2 and y with 40 in each equation.

$$y = 15x + 10$$

$$\boxed{} \overset{?}{=} 15 \left(\boxed{}\right) + 10$$

$$40 = \boxed{}$$

$$y = 20x$$

$$\boxed{} \overset{?}{=} 20 \left(\boxed{}\right)$$

$$40 = \boxed{}$$

Since both of the sentences are true, the solution of the system is (2, 40). This means that if you take lessons for 2 hours, the cost at both stores is the same, $40.

Check

Amanda pays a one-time fee of $100 and a monthly fee of $10 to belong to a gym. Maria pays only a monthly fee of $20 to belong to her gym. Write and solve a system of equations that can be used to determine when each person will pay the same amount.

Part A Select the two equations that represent the situation, where x represents the number of months and y represents the total amount paid.

(A) $y = 20x + 10x$

(B) $y = 100 + 10x$

(C) $y = 20x + 100$

(D) $y = 20x$

(E) $y = 100 + 10x + 20y$

Part B What is the solution of the system of equations?

Part C Interpret the solution.

 Show your work here

Go Online You can complete an Extra Example online.

A total of 75 loaves of banana bread and bags of popcorn were donated for a bake sale to raise money for the football team. There were four times as many bags of popcorn donated as loaves of banana bread.

Write and solve a system of equations to determine how many of each item was donated.

Part A Write a system of equations.

Words
The total number of loaves of banana bread and bags of popcorn is _____.
There were _____ times as many bags of popcorn as loaves of bread.
Variables
Let x represent the number of loaves of banana bread.
Let y represent the number of bags of popcorn.
Equations
Total items donated: $x + y = \boxed{}$
Four times as many bags of popcorn as loaves of bread: $y = \boxed{}$

Part B Solve the system of equations. Use substitution.

Since y is equal to $4x$, replace y with $4x$ in the other equation, $x + y = 75$. Then solve the equation.

$x + y = 75$ Write the equation.

$x + \boxed{} = 75$ Replace y with $4x$.

$\boxed{} = 75$ Combine like terms.

$x = \boxed{}$ Divide each side by 5.

Substitute 15 for x in either of the original equations to find the value of y.

$y = \boxed{}$

Part C Interpret the solution.

The solution is (15, 60). This means that 15 loaves of banana bread and 60 bags of popcorn were donated.

Check

Mr. Bowler cooked a total of 45 hamburgers and hot dogs at a cookout. He cooked twice as many hot dogs as hamburgers. Write and solve a system of equations that can be used to find the number of each item he cooked.

Part A Select the two equations that represent the situation, where x represents the number of hamburgers and y represents the number of hot dogs.

Ⓐ $x = 2y$

Ⓑ $y = 2x$

Ⓒ $45 + x = y$

Ⓓ $45 + y = x$

Ⓔ $x + y = 45$

Part B What is the solution of the system of equations?

 Show your work here

Part C Interpret the solution.

🐦 **Go Online**
You can complete an Extra Example online.

🌐 **Example 4** Write and Solve Systems of Equations

💭 **Think About It!**
How would you begin writing the system?

Gregory's Motorsports has motorcycles (two wheels) and ATVs (four wheels) in stock. The store has a total of 45 vehicles that, together, have 130 wheels.

Write and solve a system of equations to determine how many of each vehicle the store has in stock.

Part A Write a system of equations.

Words
The total number of motorcycles and ATVs is _____.
The number of motorcycle wheels (2) and ATV wheels (4) is _____.
Variables
Let x represent the number of motorcycles.
Let y represent the number of ATVs.
Equations
Total vehicles: $x + y = \boxed{}$
Total wheels: $2x + 4y = \boxed{}$

(continued on next page)

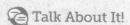
Part B Solve the system of equations.

Multiply the equation $x + y = 45$ by -2 to eliminate the variable x.

$$-2(x + y) = -2(45) \quad \rightarrow \quad -2x - 2y = -90$$
$$2x + 4y = 130 \quad \rightarrow \quad 2x + 4y = 130$$

Add the equations, eliminating the variable x. Then continue to solve the system.

$$\begin{aligned} -2x - 2y &= -90 \\ (+)\ 2x + 4y &= 130 \\ \hline 0 + 2y &= 40 \end{aligned}$$

Align like terms.

Add; the variable x is eliminated.

$y = \boxed{}$ Divide each side by 2.

$x = \boxed{}$ Find the value of x.

Part C Interpret the solution.

The solution is (25, 20). This means that the store has 25 motorcycles and 20 ATVs.

Check

The cost of 2 bagels and 1 can of orange juice is $8. The cost of 3 bagels and 2 cans of orange juice is $13. Write and solve a system of equations that can be used to find the cost of a bagel and an orange juice.

Part A Select the two equations that represent the situation, where x represents the cost of each bagel and y represents the cost of a can of orange juice.

Ⓐ $2x + y = 8$ Ⓓ $3x + 2y = 13$

Ⓑ $3 + 8x = y$ Ⓔ $13x + 2y = 3$

Ⓒ $8x + 13y = 5$

Part B What is the solution of the system of equations?

Part C Interpret the solution.

Show your work here

▶ **Go Online** You can complete an Extra Example online.

⊕ Apply Yogurt Shop

At a yogurt shop, the Miller family purchased 4 frozen yogurts and 2 smoothies and spent a total of $24. The Patel family purchased 1 frozen yogurt and 2 smoothies and spent a total of $12.75. The Reese family purchased 1 frozen yogurt and 3 smoothies. If each frozen yogurt costs the same and each smoothie costs the same, how much did the Reese family spend?

▶ Go Online
Watch the animation.

1 What is the task?

Make sure you understand exactly what question to answer or problem to solve. You may want to read the problem three times. Discuss these questions with a partner.

First Time Describe the context of the problem, in your own words.
Second Time What mathematics do you see in the problem?
Third Time What are you wondering about?

2 How can you approach the task? What strategies can you use?

Record your observations here

3 What is your solution?

Use your strategy to solve the problem.

Show your work here

💬 Talk About It!

Explain the method you used to solve the problem. Then explain how you could solve the problem another way.

4 How can you show your solution is reasonable?

✏ **Write About It!** Write an argument that can be used to defend your solution.

Check

Mr. Signet tracked the number of Calories he consumed from drinking tea and coffee over three weeks. In the first week, he drank 4 cups of tea and 5 cups of coffee, for a total of 35 Calories. In the second week, he drank 4 cups of tea and 3 cups of coffee, for a total of 25 Calories. In the third week, he drank 3 cups of tea and 2 cups of coffee. If each cup of tea contained the same number of Calories, and each cup of coffee contained the same number of Calories, how many Calories did he consume from drinking tea and coffee in the third week?

Show your work here

Go Online You can complete an Extra Example online.

Pause and Reflect

How do you know when a system of equations is needed to model a real-world situation?

Record your observations here

Practice

🔘 **Go Online** You can complete your homework online.

Write and solve a system of equations that represents each situation. Interpret the solution. (Examples 1–4)

1. The sum of two numbers is 20.5. Their difference is 6.5. Find the two numbers.

2. Tadeo volunteered at the library 6 times as many hours over the weekend as Dylan. Together, they volunteered a total of 14 hours. How many hours did each person volunteer over the weekend?

3. Tiana placed two orders for flowers and bushes. The first order was for 24 flowers and 6 bushes. The total of the first order was $144. The second order was for 18 flowers and 3 bushes. The total of the second order was $90. What is the cost of each plant?

4. Mrs. Adesso wants to take her class on a trip to either the science center or natural history museum. The science center charges $7 per student, plus $75 for a guided tour. The natural history museum charges $8 per student, plus $50 for a guided tour. For what number of students is the cost of the trip the same at each museum?

Test Practice

5. Open Response It costs $5 per hour to rent a snowboard from a certain ski rental company, plus a $50 deposit. Another ski rental company charges $10 per hour to rent a snowboard, plus a $25 deposit. For what number of hours is the cost to rent a snowboard the same at each company? What is the cost of renting a snowboard for this number of hours?

Hours, x: ☐

Cost, y: ☐

Apply

6. At a farmer's market, Amar purchased 4 jars of salsa and 3 cucumbers and spent a total of $12.25. Dylan purchased 1 jar of salsa and 2 cucumbers and spent a total of $4. Dakota purchased 1 jar of salsa and 5 cucumbers. If each jar of salsa costs the same and each cucumber costs the same, how much did Dakota spend?

7. The table shows the number of days and the total number of miles Sydney ran and cycled each week. Each day that she ran, she ran the same number of miles. Each day that she cycled, she cycled the same number of miles. Complete the table to find the total number of miles Sydney ran and cycled in Week C.

Week	Number of Days Ran	Number of Days Cycled	Total Number of Miles
A	2	3	40
B	3	4	55
C	4	2	

8. Create Write a real-world problem that can be solved using a system of equations. Then solve the problem.

9. MP **Persevere with Problems** A chemist is mixing solutions. Solution A is 15% acid, and solution B is 30% acid. She mixes the two solutions to make 12 liters of a new solution that is 25% acid. How many liters of each type of solution did the chemist use to make the new solution?

10. MP **Find the Error** A classmate wrote the system of equations to represent the problem shown at the right. Find the mistake and correct it.

A concession stand sells hot dogs and hamburgers. At the football game, 84 hot dogs and 36 hamburgers were sold for $276. At another football game, 18 hamburgers and 60 hot dogs were sold for $174. What is the cost of each hot dog and each hamburger?

$$84x + 36y = 276$$
$$18x + 60y = 174$$

📖 **Foldables** Use your Foldable to help review the module.

Solve Systems of Equations

Solve Graphically	Solve Graphically	Solve Graphically

Rate Yourself!

Complete the chart at the beginning of the module by placing a checkmark in each row that corresponds with how much you know about each topic after completing this module.

Write about one thing you learned.	Write about a question you still have.

Reflect on the Module

Use what you learned about systems of linear equations to complete the graphic organizer.

ℯ Essential Question

How can systems of equations be helpful in solving everyday problems?

Solve Systems by Graphing	Solve Systems by Substitution	Solve Systems by Elimination
How do you solve?	How do you solve?	How do you solve?
What is the solution?	What is the solution?	What is the solution?

Real-World Example

Test Practice

1. Grid Consider the system of equations.
(Lesson 1)

$y = 2x$
$y = x - 1$

A. Graph the system of equations on the coordinate plane.

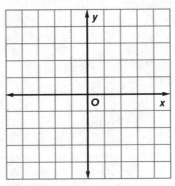

B. Name the solution of the system.

2. Open Response Consider the graph of the system of equations. What number of solutions, if any, does this system have? Justify your response. **(Lesson 1)**

3. Table Item Indicate whether each of the following systems of equations has *no solution, one solution,* or *an infinite number of solutions*. **(Lesson 2)**

Systems of Equations	Number of Solutions		
	none	one	infinite
$y = \frac{3}{5}x + 1$			
$5y = 3x + 5$			
$y = 3 - x$			
$y = -2x$			
$y = 2x + 3$			
$4y = 8x - 1$			
$2x + y = 4$			
$x - y = 1$			

4. Open Response Ramona and Malika are walking along linear routes in their county. On a map of their county, Ramona's route passes through the coordinates (2, 1) and (0, −2). Malika's route passes through the coordinates (0, 3) and (−2, 0). Do the friends pass through a common coordinate, walk along the same route, or never cross routes? Explain your answer. **(Lesson 2)**

5. Multiple Choice Which of the following is the solution of the system of equations? **(Lesson 3)**

$y = x + 6$
$y = 2x$

Ⓐ (2, 4)

Ⓑ (−2, −4)

Ⓒ (6, 12)

Ⓓ (−6, −12)

6. Multiselect Suppose you are asked to solve the following system of equations by substitution. **(Lesson 3)**

$y = 2x + 5$
$3x + 2y = −4$

Which of the following statements are accurate regarding the situation? Select all that apply.

☐ You can replace x with $2x + 5$ in the equation $3x + 2y = −4$.

☐ You can replace y with $2x + 5$ in the equation $3x + 2y = −4$.

☐ The solution of the system of equations is (−2, 1).

☐ The solution of the system of equations is (2, −1).

☐ There is no solution of the system of equations.

7. Multiple Choice Which of the following arithmetic operations would be best used to solve the system of equations by elimination? **(Lesson 4)**

$6x + 2y = 4$
$3x − 2y = 11$

Ⓐ addition

Ⓑ subtraction

Ⓒ multiplication

Ⓓ division

8. Open Response Consider the system of equations. **(Lesson 4)**

$3x + 4y = 2$
$2x − y = 5$

A. Solve the system of equations by elimination.

$$\left(\boxed{}, \boxed{}\right)$$

B. Check the solution.

9. Multiselect At Tools Plus, there is a one-time fee of $20 and an hourly fee of $10 to rent a rototiller. At Tools Unlimited, there is only an hourly fee of $15 per hour to rent a rototiller. Write and solve a system of equations that can be used to determine the number of hours for which the rental fee would be the same at both places. **(Lesson 5)**

A. Select the two equations that represent the situation, where x represents the number of hours and y represents the total amount paid.

☐ $y = 10x + 15x$

☐ $y = 15x + 20$

☐ $y = 10x + 15x + 20$

☐ $y = 10x + 20$

☐ $y = 15x$

B. What is the solution of the system of equations?

What Are Foldables and How Do I Create Them?

Foldables are three-dimensional graphic organizers that help you create study guides for each module in your book.

Step 1 Go to the back of your book to find the Foldable for the module you are currently studying. Follow the cutting and assembly instructions at the top of the page.

Step 2 Go to the Module Review at the end of the module you are currently studying. Match up the tabs and attach your Foldable to this page. Dotted tabs show where to place your Foldable. Striped tabs indicate where to tape the Foldable.

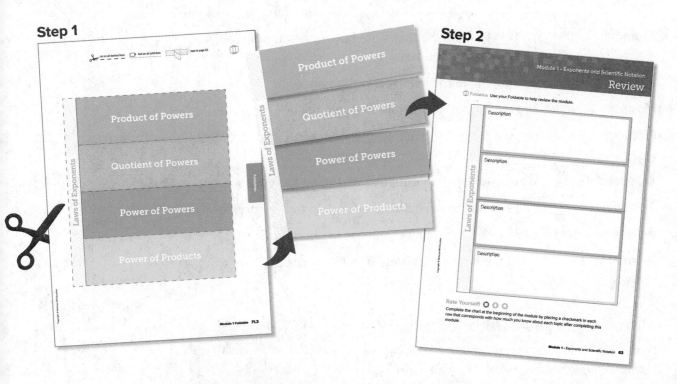

How Will I Know When to Use My Foldable?

You will be directed to work on your Foldable at the end of selected lessons. This lets you know that it is time to update it with concepts from that lesson. Once you've completed your Foldable, use it to study for the module test.

How Do I Complete My Foldable?

No two Foldables in your book will look alike. However, some will ask you to fill in similar information. Below are some of the instructions you'll see as you complete your Foldable. **HAVE FUN** learning math using Foldables!

Instructions and What They Mean

Best Used to...	Complete the sentence explaining when the concept should be used.
Definition	Write a definition in your own words.
Description	Describe the concept using words.
Equation	Write an equation that uses the concept. You may use one already in the text or you can make up your own.
Example	Write an example about the concept. You may use one already in the text or you can make up your own.
Formulas	Write a formula that uses the concept. You may use one already in the text.
How do I ...?	Explain the steps involved in the concept.
Models	Draw a model to illustrate the concept.
Picture	Draw a picture to illustrate the concept.
Solve Algebraically	Write and solve an equation that uses the concept.
Symbols	Write or use the symbols that pertain to the concept.
Write About It	Write a definition or description in your own words.
Words	Write the words that pertain to the concept.

Module 1 · Exponents and Scientific

📖 Foldables Use your Foldable to help review the module.

Rev

Description

Meet Foldables Author Dinah Zike

Dinah Zike is known for designing hands-on manipulatives that are used nationally and internationally by teachers and parents. Dinah is an explosion of energy and ideas. Her excitement and joy for learning inspires everyone she touches.

Laws of Exponents

Product of Powers

Quotient of Powers

Power of Powers

Power of Products

cut on all dashed lines

fold on all solid lines

tape to page 63

Examples

Examples

Examples

Examples

cut on all dashed lines   fold on all solid lines   tape to page 123

Real Numbers

Rational

Irrational

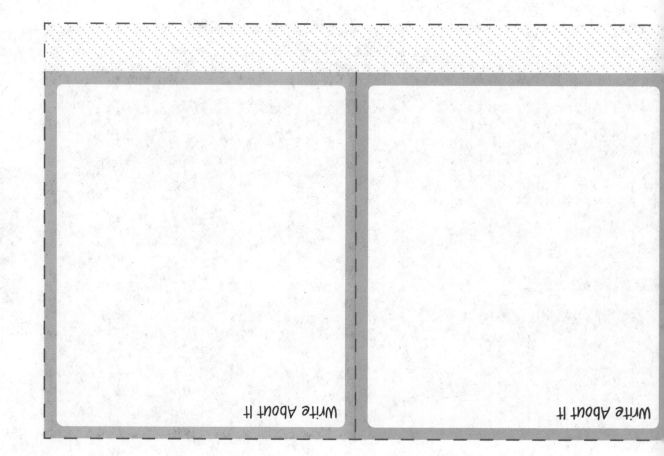

Write About It

Write About It

Solving Equations

property

solution

Step 1

Step 2

Step 3

Step 4

Distributive Property

Addition or Subtraction Property of Equality

Multiplication or Division Property of Equality

Tab 2

Tab 1

proportional linear
relationships

nonproportional linear
relationships

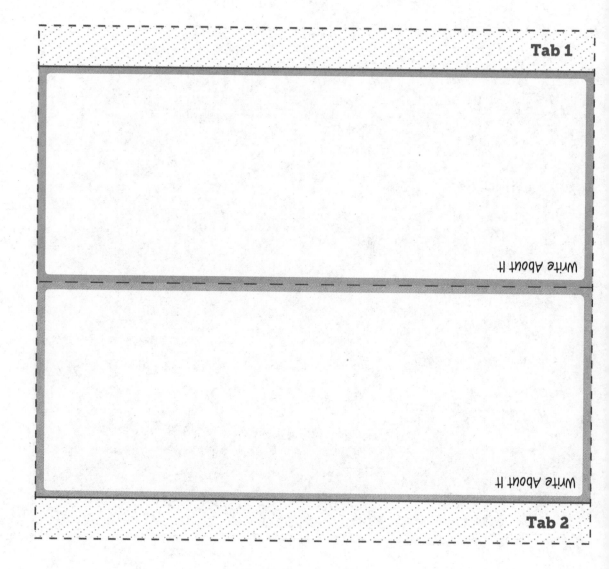

Tab 1

Write About It

Write About It

Tab 2

cut on all dashed lines — fold on all solid lines — tape to page 313

Relations and Functions

relations

functions

linear

nonlinear

Foldables

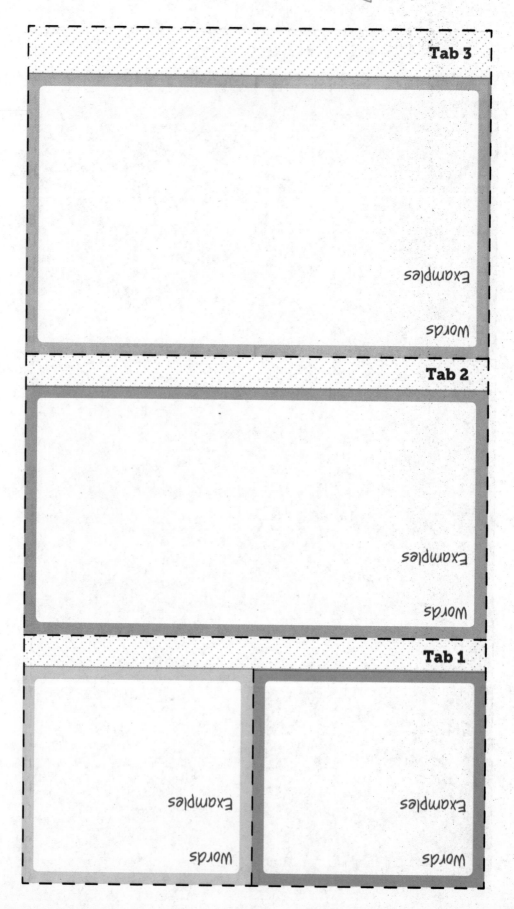

Tab 3

Examples

Words

Tab 2

Examples

Words

Tab 1

Examples

Examples

Words

Words

cut on all dashed lines | fold on all solid lines | tape to page 375

Solve Systems of Equations

one solution	no solution	infinite number of solutions

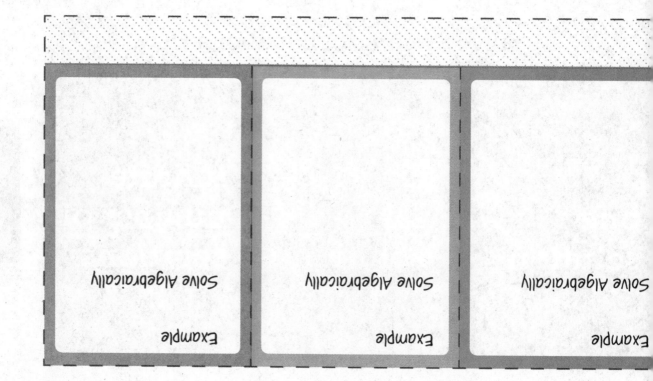

Solve Algebraically

Example

Solve Algebraically

Example

Solve Algebraically

Example

Glossary

The Multilingual eGlossary contains words and definitions in the following 14 languages:

Arabic	English	Hmong	Russian	Urdu
Bengali	French	Korean	Spanish	Vietnamese
Brazilian Portuguese	Haitian Creole	Mandarin	Tagalog	

English	Español

A

algebra (Lesson 9-1) A branch of mathematics that involves expressions with variables.

álgebra Rama de las matemáticas que trabaja con expresiones con variables.

alternate exterior angles (Lesson 7-1) Exterior angles that lie on opposite sides of the transversal.

ángulos alternos externos Ángulos externos que se encuentran en lados opuestos de la transversal.

alternate interior angles (Lesson 7-1) Interior angles that lie on opposite sides of the transversal.

ángulos alternos internos Ángulos internos que se encuentran en lados opuestos de la transversal.

angle of rotation (Lesson 8-3) The degree measure of the angle through which a figure is rotated.

ángulo de rotación Medida en grados del ángulo sobre el cual se rota una figura.

Angle-Angle Similarity (Lesson 9-4) If two angles of one triangle are congruent to two angles of another triangle, then the triangles are similar.

similitud ángulo-ángulo Si dos ángulos de un triángulo son congruentes con dos ángulos de otro triángulo, entonces los triángulos son similares.

B

bar notation (Lesson 2-1) In repeating decimals, the line or bar placed over the digits that repeat.

notación de barra Línea o barra que se coloca sobre los dígitos que se repiten en decimales periódicos.

base (Lesson 1-1) In a power, the number that is the common factor. In 10^3, the base is 10. That is, $10^3 = 10 \times 10 \times 10$.

base En una potencia, el número que es el factor común. En 10^3, la base es 10. Es decir, $10^3 = 10 \times 10 \times 10$.

bivariate data (Lesson 11-1) Data with two variables, or pairs of numerical observations.

datos bivariantes Datos con dos variables, o pares de observaciones numéricas.

C

center of dilation (Lesson 8-4) The center point from which dilations are performed.

centro de la homotecia Punto fijo en torno al cual se realizan las homotecias.

center of rotation (Lesson 8-3) A fixed point around which shapes move in a circular motion to a new position.

centro de rotación Punto fijo alrededor del cual se giran las figuras en movimiento circular alrededor de un punto fijo.

cluster (Lesson 11-1)   A collection of points that are close together in a scatter plot.

racimo   Una colección de puntos que están muy juntos en un diagrama de dispersión.

coefficient (Lesson 3-1)   The numerical factor of a term that contains a variable.

coeficiente   Factor numérico de un término que contiene una variable.

composite solid (Lesson 10-5)   An object made up of more than one solid.

sólido complejo   Cuerpo compuesto de más de un sólido.

composition of transformations (Lesson 9-1)
The resulting transformation when a transformation is applied to a figure and then another transformation is applied to its image.

composición de transformaciones   Transformación que resulta cuando se aplica una transformación a una figura y luego se le aplica otra transformación a su imagen.

cone (Lesson 10-2)   A three-dimensional figure with one circular base connected by a curved surface to a single point.

cono   Una figura tridimensional con una circular base conectada por una superficie curva para un solo punto.

congruent (Lesson 9-1)   Having the same measure; if one image can be obtained from another by a sequence of rotations, reflections, or translations.

congruente   Que tienen la misma medida; si una imagen puede obtenerse de otra por una secuencia de rotaciones, reflexiones o traslaciones.

constant (Lesson 3-1)   A term without a variable.

constante   Término sin variables.

constant of proportionality (Lesson 4-4)   The constant ratio in a proportional linear relationship.

constante de proporcionalidad   La razón constante en una relación lineal proporcional.

constant of variation (Lesson 4-4)   A constant ratio in a direct variation.

constante de variación   Razón constante en una relación de variación directa.

constant rate of change (Lesson 4-1)   The rate of change between any two points in a linear relationship is the same or *constant*.

tasa constante de cambio   La tasa de cambio entre dos puntos cualesquiera en una relación lineal permanece igual o *constante*.

converse (Lesson 7-4)   The converse of a theorem is formed when the *if* and *then* statements are reversed.

recíproco   El recíproco de un teorema se forma cuando las declaraciones de *si* y *luego* se invierten.

converse of the Pythagorean Theorem (Lesson 7-4)
A theorem that can be used to test whether a triangle is a right triangle. If the sides of the triangle have lengths a, b, and c, such that $c^2 = a^2 + b^2$, then the triangle is a right triangle.

el recíproco del teorema de Pitágoras   Un teorema que puede usarse para probar si un triángulo es un triángulo rectángulo. Si los lados del triángulo tienen longitudes a, b, y c, tales que $c^2 = a^2 + b^2$, entonces el triángulo es un triángulo rectángulo.

coordinate plane (Lesson 4-1)   A coordinate system in which a horizontal number line and a vertical number line intersect at their zero points.

plano de coordenadas   Sistema de coordenadas en que una recta numérica horizontal y una recta numérica vertical se intersecan en sus puntos cero.

corresponding angles (Lesson 7-1)   Angles that are in the same position on two parallel lines in relation to a transversal.

ángulos correspondientes   Ángulos que están en la misma posición sobre dos rectas paralelas en relación con la transversal.

corresponding parts (Lesson 4-3)   Parts of congruent or similar figures that are in the same relative position.

partes correspondientes   Partes de figuras congruentes o similares que están en la misma posición relativa.

counterexample (Lesson 2-3)   A statement or example that shows a conjecture is false.

contraejemplo   Ejemplo o enunciado que demuestra que una conjetura es falsa.

cube root (Lesson 2-2)   One of three equal factors of a number. If $a^3 = b$, then a is the cube root of b. The cube root of 64 is 4 since $4^3 = 64$.

raíz cúbica   Uno de tres factores iguales de un número. Si $a^3 = b$, entonces a es la raíz cúbica de b. La raíz cúbica de 64 es 4, dado que $4^3 = 64$.

cylinder (Lesson 10-1)   A three-dimensional figure with two parallel congruent circular bases connected by a curved surface.

cilindro   Una figura tridimensional con dos paralelas congruentes circulares bases conectados por una superficie curva.

D

dilation (Lesson 8-4)   A transformation that enlarges or reduces a figure by a scale factor.

homotecia   Transformación que produce la ampliación o reducción de una imagen por un factor de escala.

direct variation (Lesson 4-4)   A relationship between two variable quantities with a constant ratio. A proportional linear relationship.

variación directa   Relación entre dos cantidades variables con una razón constante. Una relación lineal proporcional.

E

elimination (Lesson 6-4)   An algebraic method that can be used to find the exact solution of a system of equations by eliminating one of the variables.

eliminación   Un método algebraico que se puede usar para encontrar la solución exacta de un sistema de ecuaciones mediante la eliminación de una de las variables.

evaluate (Lesson 1-1)   To find the value of an expression.

evaluar   Calcular el valor de una expresión.

exponent (Lesson 1-1)   In a power, the number of times the base is used as a factor. In 10^3, the exponent is 3.

exponente   En una potencia, el número de veces que la base se usa como factor. En 10^3, el exponente es 3.

exterior angle (Lesson 7-2)   An angle between one side of a polygon and the extension of an adjacent side.

ángulo exterior   Un ángulo entre un lado de un polígono y la extensión de un lado adyacente.

exterior angles (Lesson 7-1)   The four outer angles formed by two lines cut by a transversal.

ángulo externo   Los cuatro ángulos exteriores que se forman cuando una transversal corta dos rectas.

F

function (Lesson 5-1)   A relation in which each member of the input is paired with exactly one member of the output.

función   Relación en la que cada elemento de la entrada le corresponde exactamente con un único elemento de la salida.

function table (Lesson 5-2)   A table organizing the input, rule, and output of a function.

tabla de funciones   Tabla que organiza la regla de entrada y de salida de una función.

H

hemisphere **(Lesson 10-3)** One of two congruent halves of a sphere.

hypotenuse **(Lesson 7-3)** The side opposite the right angle in a right triangle.

hemisferio Una de dos mitades congruentes de una esfera.

hipotenusa El lado opuesto al ángulo recto de un triángulo rectángulo.

I

image **(Lesson 8-1)** The resulting figure after a transformation.

imagen Figura que resulta después de una transformación.

indirect measurement **(Lesson 9-5)** A technique using properties of similar polygons to find distances or lengths that are difficult to measure directly.

medición indirecta Técnica que usa las propiedades de polígonos semejantes para calcular distancias o longitudes difíciles de medir directamente.

initial value **(Lesson 4-5)** The starting value in a real-world situation in which an equation can be written. The *y*-intercept of a linear function.

valor inicial El valor inicial en una situación real en la que se puede escribir una ecuación. La intersección *y* de una función lineal.

input **(Lesson 5-1)** The set of *x*-coordinates in a relation.

entrada El conjunto de *x*-coordenadas en una relación.

integers **(Lesson 2-1)** The set of whole numbers and their opposites.

enteros El conjunto de números enteros y sus opuestos.

interior angle **(Lesson 7-2)** An angle inside a polygon.

ángulo interno Ángulo dentro de un polígono.

interior angles **(Lesson 7-1)** The four inside angles formed by two lines cut by a transversal.

ángulo interno Los cuatro ángulos internos formados por dos rectas intersecadas por una transversal.

inverse operations **(Lesson 2-2)** Pairs of operations that undo each other. Addition and subtraction are inverse operations. Multiplication and division are inverse operations.

peraciones inversas Pares de operaciones que se anulan mutuamente. La adición y la sustracción son operaciones inversas. La multiplicación y la división son operaciones inversas.

irrational number **(Lesson 2-3)** A number that cannot be expressed as the ratio $\frac{a}{b}$, where a and b are integers and $b \neq 0$.

números irracionales Número que no se puede expresar como la proporción $\frac{a}{b}$, donde a y b son enteros y $b \neq 0$.

L

legs **(Lesson 7-3)** The two sides of a right triangle that form the right angle.

catetos Los dos lados de un triángulo rectángulo que forman el ángulo recto.

like terms (Lesson 3-3) Terms that contain the same variable(s) to the same powers.

términos semejantes Términos que contienen la misma variable o variables elevadas a la misma potencia.

linear (Lesson 4-1) To fall in a straight line.

lineal Que cae en una línea recta.

linear equation (Lesson 4-1) An equation with a graph that is a straight line.

ecuación lineal Ecuación cuya gráfica es una recta.

linear function (Lesson 5-2) A function in which the graph of the solutions forms a line.

función lineal Función en la cual la gráfica de las soluciones forma una recta.

linear relationship (Lesson 4-1) A relationship that has a straight-line graph.

relación lineal Relación cuya gráfica es una recta.

line of fit (Lesson 11-2) A line that is very close to most of the data points in a scatter plot.

línea de ajuste Línea que más se acerca a la mayoría de puntos de los datos en un diagrama de dispersión.

line of reflection (Lesson 8-2) The line over which a figure is reflected in a transformation.

línea de reflexión Línea a través de la cual se refleja una figura en una transformación.

line segment (Lesson 7-2) Part of a line containing two endpoints and all of the points between them.

segmento de línea Parte de una línea que contiene dos extremos y todos los puntos entre ellos.

M

monomial (Lesson 1-2) A number, a variable, or a product of a number and one or more variables.

monomio Un número, una variable o el producto de un número por una o más variables.

N

natural numbers (Lesson 2-1) The set of numbers used for counting.

números naturales El conjunto de números utilizado para el recuento.

negative exponent (Lesson 1-4) The result of repeated division used to represent very small numbers.

exponente negative El resultado de la división repetida se utiliza para representar números muy pequeños.

nonlinear function (Lesson 5-5) A function whose rate of change is not constant. The graph of a nonlinear function is not a straight line.

función no lineal Función cuya tasa de cambio no es constante. La gráfica de una función no lineal no es una recta.

Order of Operations (Lesson 1-1) When evaluating expressions with powers or more than one operation, use these rules.

1. Simplify the expression inside the grouping symbols.

2. Evaluate all powers.

3. Perform multiplication and division in order from left to right.

4. Perform addition and subtraction in order from left to right.

orden de operaciones Al evaluar expresiones con poderes o más de una operación, utilice estas reglas.

1. Simplifique la expresión dentro de los símbolos de agrupación.

2. Evaluar todos los poderes.

3. Realizar multiplicación y división en orden de izquierda a derecha.

4. Realizar la suma y resta en orden de izquierda a derecha.

ordered pair (Lesson 4-1) A pair of numbers used to locate a point in the coordinate plane. The ordered pair is written in this form: (x-coordinate, y-coordinate).

par ordenado Par de números que se utiliza para ubicar un punto en un plano de coordenadas. Se escribe de la siguiente forma: (coordenada x, coordenada y).

origin (Lesson 4-1) The point of intersection of the x-axis and y-axis in a coordinate plane.

origen Punto en que el eje x y el eje y se intersecan en un plano de coordenadas.

outlier (Lesson 11-1) A data point that is distinctly separate from the rest of the data.

valor atípico Un punto de datos que está claramente separado del resto de los datos.

output (Lesson 5-1) The set of y-coordinates in a relation.

salida El conjunto de y-coordenadas en una relación.

parallel lines (Lesson 7-1) Lines in the same plane that never intersect or cross. The symbol ‖ means parallel.

rectas paralelas Rectas que yacen en un mismo plano y que no se intersecan. El símbolo ‖ significa paralela a.

perfect cube (Lesson 2-2) A number whose cube root is an integer. 27 is a perfect cube because its cube root is 3.

cubo perfecto Número cuya raíz cúbica es un número entero. 27 es un cubo perfecto porque su raíz cúbica es 3.

perfect square (Lesson 2-2) A number whose square root is a whole number. 25 is a perfect square because its square root is 5.

cuadrados perfectos Número cuya raíz cuadrada es un número entero. 25 es un cuadrado perfecto porque su raíz cuadrada es 5.

perpendicular lines (Lesson 7-1) Two lines that intersect to form right angles.

rectas perpendiculares Dos rectas que se intersecan formando ángulos rectos.

power (Lesson 1-1) A product of repeated factors using an exponent and a base. The power 7^3 is read *seven to the third power,* or *seven cubed.*

potencia Producto de factores repetidos con un exponente y una base. La potencia 7^3 se lee *siete a la tercera potencia* o *siete al cubo.*

Power of a Power Property (Lesson 1-3) A property that states to find the power of a power, multiply the exponents.

Power of a Product Property (Lesson 1-3) A property that states to find the power of a product, find the power of each factor and multiply.

preimage (Lesson 8-1) The original figure before a transformation.

principal square root (Lesson 2-2) The positive square root of a number.

Product of Powers Property (Lesson 1-2) A property that states to multiply powers with the same base, add their exponents.

proof (Lesson 7-4) A logical argument in which each statement that is made is supported by a statement that is accepted as true.

Pythagorean Theorem (Lesson 7-3) In a right triangle, the square of the length of the hypotenuse c is equal to the sum of the squares of the lengths of the legs a and b. $a^2 + b^2 = c^2$

potencia de una propiedad de potencia Una propiedad que declara encontrar el poder de un poder, multiplicar los exponentes.

potencia de una propiedad de producto Una propiedad que declara encontrar el poder de un producto, encuentra el poder de cada factor y se multiplica.

preimagen Figura original antes de una transformación.

raíz cuadrada principal La raíz cuadrada positiva de un número.

producto de la propiedad de los poderes Una propiedad que declara multiplicar poderes con la misma base, añade sus exponentes.

prueba Argumento lógico en el cual cada enunciado hecho se respalda con un enunciado que se acepta como verdadero.

Teorema de Pitágoras En un triángulo rectángulo, el cuadrado de la longitud de la hipotenusa es igual a la suma de los cuadrados de las longitudes de los catetos. $a^2 + b^2 = c^2$

Q

quadrants (Lesson 4-1) The four sections of the coordinate plane.

qualitative graph (Lesson 5-6) A graph used to represent situations that do not necessarily have numerical values.

Quotient of Powers Property (Lesson 1-2) A property that states to divide powers with the same base, subtract their exponents.

cuadrantes Las cuatro secciones del plano de coordenadas.

gráfica cualitativa Gráfica que se usa para representar situaciones que no tienen valores numéricos necesariamente.

propiedad del cociente de poderes Una propiedad que declara dividir poderes con la misma base, resta sus exponentes.

R

radical sign (Lesson 2-2) The symbol used to indicate a positive square root, $\sqrt{}$.

rate of change (Lesson 4-1) A rate that describes how one quantity changes in relation to another quantity.

signo radical Símbolo que se usa para indicar una raíz cuadrada no positiva, $\sqrt{}$.

tasa de cambio Una tasa que describe cómo cambia una cantidad en relación con otra cantidad.

rational numbers (Lesson 2-1) Numbers that can be written as the ratio of two integers in which the denominator is not zero. All integers, fractions, mixed numbers, and percents are rational numbers.

números racionales Números que pueden escribirse como la razón de dos enteros en los que el denominador no es cero. Todos los enteros, fracciones, números mixtos y porcentajes son números racionales.

real numbers (Lesson 2-3) The set of rational numbers together with the set of irrational numbers.

números reales El conjunto de números racionales junto con el conjunto de números irracionales.

reflection (Lesson 8-2) A transformation where a figure is flipped over a line. Also called a flip.

reflexión Transformación en la cual una figura se voltea sobre una recta. También se conoce como simetría de espejo.

relation (Lesson 5-1) Any set of ordered pairs.

relación Cualquier conjunto de pares ordenados.

relative frequency (Lesson 11-4) The ratio of the number of experimental successes to the total number of experimental attempts.

frecuencia relativa Razón del número de éxitos experimentales al número total de intentos experimentales.

remote interior angles (Lesson 7-2) The angles of a triangle that are not adjacent to a given exterior angle.

ángulos internos no adyacentes Ángulos de un triángulo que no son adya centes a un ángulo exterior dado.

repeating decimal (Lesson 2-1) A decimal in which 1 or more digits repeat.

decimal periódico Un decimal en el que se repiten 1 o más dígitos.

right triangle (Lesson 7-3) A triangle with one right angle.

triángulo rectángulo Triángulo con un ángulo recto.

rise (Lesson 4-2) The vertical change between any two points on a line.

elevación El cambio vertical entre cualquier par de puntos en una recta.

rotation (Lesson 8-3) A transformation in which a figure is turned about a fixed point.

rotación Transformación en la cual una figura se gira alrededor de un punto fijo.

run (Lesson 4-2) The horizontal change between any two points on a line.

carrera El cambio horizontal entre cualquier par de puntos en una recta.

S

scale factor (Lesson 8-4) The ratio of the lengths of two corresponding sides of two similar polygons.

factor de escala La razón de las longitudes de dos lados correspondientes de dos polígonos semejantes.

scatter plot (Lesson 11-1) A graph that shows the relationship between bivariate data graphed as ordered pairs on a coordinate plane.

diagrama de dispersión Gráfica que muestra la relación entre datos bivariados graficadas como pares ordenados en un plano de coordenadas.

scientific notation (Lesson 1-5) A compact way of writing numbers with absolute values that are very large or very small. In scientific notation, 5,500 is 5.5×10^3.

notación científica Manera abreviada de escribir números con valores absolutos que son muy grandes o muy pequeños. En notación científica, 5,500 es 5.5×10^3.

similar (Lesson 9-3) If one image can be obtained from another by a sequence of transformations and dilations.

similar Si una imagen puede obtenerse de otra mediante una secuencia de transformaciones y dilataciones.

similar figures (Lesson 4-3) Figures that have the same shape but not necessarily the same size.

figuras semejantes Figuras que tienen la misma forma, pero no necesariamente el mismo tamaño.

slope (Lesson 4-1) The rate of change between any two points on a line. The ratio of the rise, or vertical change, to the run, or horizontal change.

pendiente Razón de cambio entre cualquier par de puntos en una recta. La razón de la altura, o cambio vertical, a la carrera, o cambio horizontal.

slope-intercept form (Lesson 4-5) An equation written in the form $y = mx + b$, where m is the slope and b is the y-intercept.

forma pendiente intersección Ecuación de la forma $y = mx + b$, donde m es la pendiente y b es la intersección y.

slope triangles (Lesson 4-3) Right triangles that fall on the same line on the coordinate plane.

triángulos de pendiente Triángulos rectos que caen en la misma línea en el plano de coordenadas.

solution (of an equation) (Lesson 4-1) A value that makes an equation true.

solución (de una ecuación) Un valor que hace que una ecuación sea verdadera.

solution (of a system of equations) (Lesson 6-1) An ordered pair that is a solution of both equations.

solución (de un sistema de ecuaciones) Un par ordenado que es una solución de ambas ecuaciones.

solid (Lesson 10-1) A three-dimensional figure formed by intersecting planes.

sólido Figura tridimensional formada por planos que se intersecan.

sphere (Lesson 10-3) The set of all points in space that are a given distance from a given point called the center.

esfera Conjunto de todos los puntos en el espacio que están a una distancia dada de un punto dado llamado centro.

square root (Lesson 2-2) One of the two equal factors of a number. If $a^2 = b$, then a is the square root of b. A square root of 144 is 12 since $12^2 = 144$.

raíz cuadrada Uno de dos factores iguales de un número. Si $a^2 = b$, la a es la raíz cuadrada de b. Una raíz cuadrada de 144 es 12 porque $12^2 = 144$.

standard form (Lesson 1-5) Numbers written without exponents.

forma estándar Números escritos sin exponentes.

substitution (Lesson 6-3) An algebraic model that can be used to find the exact solution of a system of equations.

sustitución Modelo algebraico que se puede usar para calcular la solución exacta de un sistema de ecuaciones.

system of equations (Lesson 6-1) A set of two or more equations with the same variables.

sistema de ecuaciones Sistema de ecuaciones con las mismas variables.

term **(Lesson 1-2)** Each part of an algebraic expression separated by an addition or subtraction sign.

término Cada parte de un expresión algebraica separada por un signo adición o un signo sustracción.

terminating decimal **(Lesson 2-1)** A decimal where the repeating digit is zero.

decimal finito Un decimal donde el dígito que se repite es cero.

theorem **(Lesson 7-3)** A statement or conjecture that can be proven.

teorema Un enunciado o conjetura que puede probarse.

transformation **(Lesson 8-1)** An operation that maps a geometric figure, preimage, onto a new figure, image.

transformación Operación que convierte una figura geométrica, la pre-imagen, en una figura nueva, la imagen.

translation **(Lesson 8-1)** A transformation that slides a figure from one position to another without turning.

traslación Transformación en la cual una figura se desliza de una posición a otra sin hacerla girar.

transversal **(Lesson 7-1)** A line that intersects two or more other lines.

transversal Recta que interseca dos o más rectas.

triangle **(Lesson 7-2)** A figure formed by three line segments that intersect only at their endpoints.

triángulo Figura formada por tres segmentos de recta que se intersecan sólo en sus extremos.

truncating **(Lesson 2-4)** A process of approximating a decimal number by eliminating all decimal places past a certain point without rounding.

truncando Proceso de aproximación de un número decimal eliminando todos los decimales más allá de un cierto punto sin redondear.

two-way table **(Lesson 11-4)** A table that shows data from one sample that pertain to two different categories.

tabla de doble entrada Una tabla que muestra datos de una muestra que pertenecen a dos categorías diferentes.

unit rate **(Lesson 4-1)** A rate in which the first quantity is compared to 1 unit of the second quantity.

tasa unitaria Una tasa en la que la primera cantidad se compara con 1 unidad de la segunda cantidad.

unit ratio **(Lesson 4-1)** A ratio in which the first quantity is compared to every 1 unit of the second quantity.

razón unitaria Una relación en la que la primera cantidad se compara con cada 1 unidad de la segunda cantidad.

variable **(Lesson 3-1)** A symbol, usually a letter, used to represent a number in mathematical expressions or sentences.

variable Un símbolo, por lo general, una letra, que se usa para representar números en expresiones o enunciados matemáticos.

vertex **(Lesson 7-2)** The point where the sides of an angle meet.

vértice Punto donde se encuentran los lados.

vertical line test **(Lesson 5-1)** A test used to see whether a graph is a function. If, for each value of x, a vertical line passes through no more than one point on the graph, then the graph represents a function. If the line passes through more than one point on the graph, it is not a function.

prueba de línea vertical Una prueba utilizada para ver si un gráfico es una función. Si, para cada valor de x, una línea vertical no pasa más de un punto en el gráfico, entonces el gráfico representa una función. Si la línea pasa por más de un punto en el gráfico, no es una función.

volume **(Lesson 10-1)** The measure of the space occupied by a solid. Standard units of measure are cubic units such as in^3 or ft^3.

volumen Medida del espacio que ocupa un sólido. Unidades de medida estándar son unidades cúbicas tales como $pulg^3$ o $pies^3$.

W

whole numbers **(Lesson 2-1)** The set of numbers used for counting and zero.

números enteros El conjunto de números utilizado para contar y cero.

X

x-axis **(Lesson 4-1)** The horizontal number line that helps to form the coordinate plane.

eje x La recta numérica horizontal que ayuda a formar el plano de coordenadas.

x-coordinate **(Lesson 4-1)** The first number of an ordered pair.

coordenada x El primer número de un par ordenado.

x-intercept **(Lesson 4-1)** The x-coordinate of the point where the line crosses the x-axis.

intersección x La coordenada x del punto donde cruza la gráfica el eje x.

Y

y-axis **(Lesson 4-1)** The vertical number line that helps to form the coordinate plane.

eje y La recta numérica vertical que ayuda a formar el plano de coordenadas.

y-coordinate **(Lesson 4-1)** The second number of an ordered pair.

coordenada y El segundo número de un par ordenado.

y-intercept **(Lesson 4-5)** The y-coordinate of the point where the line crosses the y-axis.

intersección y La coordenada y del punto donde cruza la gráfica el eje y.

Z

Zero Exponent Rule **(Lesson 1-4)** A rule that states that any nonzero number to the zero power is equivalent to 1.

regla de exponente cero Una regla que establece que cualquier número diferente de cero a la potencia cero es equivalente a 1.

Index

Index

Lesson 1-1 Powers and Exponents, Practice Pages 11–12

1. $(-7)^2 \cdot 5^4$ **3.** 65 **5.** $7\frac{7}{16}$ **7.** >
9. $\left(-\frac{4}{5}\right)^3$; $3^5 - 10^4$; $(9.8)^2 - 10^2$ **11.** 8,200 lakes
13. Sample answer: He substituted -4 for x, but should have taken -4 to the second power, not just 4. Placing parentheses around -4 would have helped him to take the correct value to the second power. The correct value of the expression is 729. **15.** It is incorrect. Because $4^2 = 2^4$, the first set of expressions are equal. This is not the case for $\left(\frac{1}{3}\right)^2$ and $\left(\frac{1}{2}\right)^3$ because $\left(\frac{1}{3}\right)^2 = \frac{1}{9}$ and $\left(\frac{1}{2}\right)^3 = \frac{1}{8}$.

Lesson 1-2 Multiply and Divide Monomials, Practice Pages 23–24

1. 3^9 or 19,683 **3.** $24m^4n^5$ **5.** b^7 **7.** 10^3 or 1,000 **9.** a^2c^5 **11.** Dish B **13.** 4^{16}; Sample answer: *four times* 4^{15} translates to $4 \cdot 4^{15}$ which simplifies to 4^{16}. **15.** $n = 5$

Lesson 1-3 Powers of Monomials, Practice Pages 31–32

1. 7^6 or 117,649 **3.** d^{42} **5.** $64m^{30}$
7. $-243w^{15}z^{40}$ **9.** $(6^2)^3$; $(6^2)^3 = 46,656$ and $46,656 > 1,000$ **11.** $49x^6y^{10}$ tiles
13. Sample answer: Using the Power of a Power law of exponents, both expressions simplify to 4^8. Multiplication is commutative, so $2 \cdot 4$ is the same as $4 \cdot 2$. **15.** $[(-4)^4]^5$; Sample answer: Using the Power of a Power property, $[(-4)^4]^5$ simplifies to $(-4)^{20}$, a positive number. The expression $-[(4^{12})^3]$ simplifies to $-(4^{36})$, a negative number.

Lesson 1-4 Zero and Negative Exponents, Practice Pages 41–42

1. 1 **3.** $\frac{1}{8^4}$ **5.** d^{-6} **7.** $\frac{1}{81}$ **9.** x^4 **11.** 1
13. 10^{12} times larger **15.** 5^{-7}, 5^0, 5^4; Sample answer: Written with a positive exponent, 5^{-7} is $\frac{1}{5^7}$ and is less than 1, 5^0 equals 1, and 5^4 is greater than 1. **17.** They are equivalent; Sample answer: Both expressions simplify to 1.

Lesson 1-5 Scientific Notation, Practice Pages 53–54

1. 1,600 **3.** 0.0000083 **5.** 2.204×10^9
7. 5 kilometers; Sample answer: The number 5×10^6 millimeters is unnecessarily large and would be difficult to visualize. Choosing the larger unit of measure is more appropriate. **9.** about 2×10^{-3} inch **11.** 1,089,822 insects **13.** 1.5×10^5 **15.** Sample answer: Katrina did not take into account place value and the powers of ten. The number 3.5×10^4, in standard form, is 35,000. The number 2.1×10^6 is 2,100,000. Therefore, 2.1×10^6 is greater.

Lesson 1-6 Compute with Scientific Notation, Practice Pages 61–62

1. 3×10^{22} **3.** 1.2913×10^5 **5.** about 2.72×10^8 seconds **7.** 9.6×10^{-4} gram
9. Sample answer: Each number has a factor that is a power of 10. Since the bases are the same, these properties can be applied to multiply or divide the powers of 10.
11. Sample answer: He found $5.78 \div 2$ as 2.89, but incorrectly found the power of ten. The numerator is 10^5 and the denominator is 10^{-6}. When dividing, the resulting power of ten is 10^{11}. The correct answer should be 2.89×10^{11}.

Module 1 Review Pages 65–66

1. B **3.** Harrisburg, Liberty Crossing, Glenview
5.

	true	false
$(d^4)^3 = d^{12}$	X	
$(n^5)^{10} = n^{15}$		X
$(k^7)^3 = k^{21}$	X	

7. 6^{-4} **9.** A **11.** 0.0072 **13.** 7.75×10^7

Lesson 2-1 Terminating and Repeating Decimals, Practice Pages 77–78

1. −0.6875; terminating **3.** 4.375; terminating
5. $\frac{8}{9}$ **7.** $-1\frac{5}{9}$ **9.** $0.\overline{5}$ **11.** $5\frac{2}{3}$ **13.** Milica
would save about 517 shots on goal out of
600 shots. **15.** Sample answer: $0.\overline{268}$; This
decimal written as a fraction is $\frac{268}{999}$. Because
it can be written in fraction form, it is a rational
number. **17.** Sample answer: I predict that $\frac{3}{162}$
is a non-terminating decimal, because 162 is
divisible by 9, and fractions written in
simplest form with a denominator of 9 are
non-terminating when written in decimal
form. $\frac{3}{162}$ written in decimal form is $0.0\overline{185}$,
which is a non-terminating decimal.

Lesson 2-2 Roots, Practice Pages 89–90

1. 19 **3.** $-\frac{3}{4}$ **5.** ±0.2 **7.** −8 **9.** 14 plants
11. 24 feet **13.** Sample answer: The rational
number, 2, when cubed, results in 8. However,
there is not a rational number that, when
multiplied by itself, results in 8. **15.** 9,261

Lesson 2-3 Real Numbers, Practice Pages 99–100

1. irrational **3.** rational **5.** rational
7. irrational **9.** A, C, D, E **11.** A, C **13.** true;
Sample answer: All integers can be expressed
as a ratio $\frac{a}{b}$, where a and b are integers and

$b \neq 0$, which is the definition of a rational
number. Therefore, all integers are rational
numbers. **15.** Sample answer: I would use
a calculator to find $\sqrt{8}$. The calculator shows
2.82842712474619. Then I would multiply that
answer by itself, without using the x^2 button.
If the solution is 8, then it is a terminating
decimal. If the solution is not 8, then I know
that the original solution was rounded by
the calculator, and it is not a terminating
decimal. **17.** false; The definition of a rational
number is a number expressed as a ratio $\frac{a}{b}$,
where a and b are integers and $b \neq 0$. In the
ratio $\frac{\sqrt{2}}{1}$, $\sqrt{2}$ is not an integer, therefore, it does
not satisfy the definition of a rational number.

Lesson 2-4 Estimate Irrational Numbers, Practice Pages 109–110

1. 11 **3.** 4 **5.** 17.2 **7.** 3.3 **9.** 21.2 miles per
hour **11.** 22.1 **13.** about 12 feet **15.** Sample
answer: To write the exact value for the square
root of a non-perfect square, such as $\sqrt{13}$, I
would leave it written as $\sqrt{13}$. The decimal
form would have to be rounded, no matter
how many decimal places I wrote. **17.** Sample
answer: In the same way I can estimate square
roots and cube roots, I could find the nearest
fourth root of 20. Since $16 < 20 < 81$, the fourth
root is between 2 and 3. Since 20 is closer to
16, the fourth root of 20 is about 2.

Lesson 2-5 Compare and Order Real Numbers, Practice Pages 121–122

1. <

3. <

5. $\{\pi, \frac{10}{3}, 3\frac{1}{2}, \sqrt{13}\}$

7. B **9.** 0.3 second **11.** π; Sample answer: 3.14 can be extended to 3.14000... . The number π written as a decimal is 3.141... and 3.141 > 3.140. **13.** Sample answer: 1.7 and $\sqrt{3}$; $1.7 < \sqrt{3}$

Module 2 Review Pages 125–126

1A. Varsity: $0.\overline{72}$; not terminating; Junior Varsity: 0.7; terminating **1B.** Varsity **3.** 16 **5A.** 50 **5B.** Sample answer: Let x = side length of each square. Solve the equation $5x^2 = 500$; $x = 10$. Height of shelving unit = 10×5, or 50 inches.

7.

	Rational	Irrational
-6	X	
$\sqrt{7}$		X
$3\frac{1}{2}$	X	

9.

	7	8	9
$\sqrt{70}$		X	
$\sqrt{79}$			X
$\sqrt{88}$			X
$\sqrt{52}$	X		
$\sqrt{60}$		X	
$\sqrt{47}$	X		
$\sqrt{65}$		X	

11A. >

11B.

Lesson 3-1 Solve Equations with Variables on Each Side, Practice Pages 135–136

1. -3 **3.** 12 **5.** -1.2 **7.** -3 **9.** 5 hours **11.** Sample answer: You can multiply each side of the equation by the least common denominator, 10, using the Distributive Property. Doing so will eliminate the fractional coefficients. **13.** false; Sample answer: You can also solve the equation by first adding or subtracting the constants to both sides of the equation.

Lesson 3-2 Write and Solve Equations with Variables on Each Side, Practice Pages 143–144

1. Let m = the number of months; $45 + 4m = 61 + 2m$; 8 months **3.** Let p = the number of pounds; $14 + 2.25p = 20 + 1.50p$; 8 pounds **5.** $15 - 0.75g = 13 - 0.50g$; 8 games **7.** $780 **9.** Let m = the number of hours worked on Monday; $m + (m + 3) + (2m + 1) = 5m + 2$; 2 hours

Lesson 3-3 Solve Multi-Step Equations, Practice Pages 149–150

1. -2 **3.** 4.5 **5.** 4 **7.** -5 **9.** $x = 1$; 3 units **11.** Sample answer: The Distributive Property allows you to expand the expressions that contain grouping symbols. Then you can combine any like terms and solve the equation using the properties of equality.

Lesson 3-4 Write and Solve Multi-Step Equations, Practice Pages 157–158

1. Let ℓ = the length; $86\frac{1}{2} = 2\ell + 2\left(2\ell - 40\frac{3}{4}\right)$; 28 feet **3.** Equation: $5(t + 3.5) = 3(2t + 2.75)$; Cost of a Ticket: $9.25 **5.** Sample answer: Andrea bought 4 identical snacks and 15 batting cage tokens. Each token costs $2.25 less than each snack. She spent a total of $18.50. What was the cost of each snack? $4x + 15(x - 2.25) = 18.50$; $2.75 **7.** Sample answer: She did not include Petra and Valentina in their groups. The correct equation is $5(x + 5.5) = 4(1.5x + 4.75)$.

Lesson 3-5 Determine the Number of Solutions, Practice Pages 167–168

1. no solution   **3.** infinitely many solutions
5. $-5x - 70$ **7.** Sample constant: $-12x + 6$
9. D   **11.** Fatima   **13.** Sample answer: The solution $x = 0$ means that 0 is the solution to the equation and the equation has one solution, 0.   **15.** Sample answer: one solution: $-4x + 7 = -5x + 9$; no solution: $-4x + 7 = -4x + 8$; infinite solutions: $-4x + 7 = -4x + 7$

Module 3 Review Pages 171–172

1. 4   **3A.** $b = -9$
3B. $\frac{2}{3}b + 5 = -\frac{1}{3}b - 4$
$\frac{2}{3}(-9) + 5 \overset{?}{=} -\frac{1}{3}(-9) - 4$
$-6 + 5 \overset{?}{=} 3 - 4$
$-1 = -1$
5A. $5m + 20 = 7m$ **5B.** $m = 10$ **7.** -3
9. $x = 2$

11.

	One Solution	No Solution	Infinitely Many Solutions
$4(x + 8) = 2(4 + 2x)$		X	
$3(2x + 1) = 3 + 6x$			X
$2(x + 5) = 5x + 1$	X		

Lesson 4-1 Proportional Relationships and Slope, Practice Pages 189–190

1. The slope of the line is $\frac{333.3\overline{3}}{1}$ or $333.3\overline{3}$. This means the book sales were about \$333.33 each day. The unit rate is about \$333.33 per day, which is the same as the slope.   **3.** The slope of the line is $\frac{2.3}{1}$ or 2.3. This means that the movie grossed \$2.3 million each week.

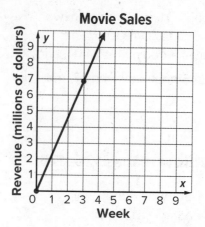

Movie Sales

5. Craig; Sample answer: The unit rate of Craig's day trips is 21.8 miles per day. Since Rei biked 23.6 miles per day, and $23.6 > 21.8$, Craig biked the lower number of miles each day.   **7.** Sample answer: The unit rate of a proportional relationship is the comparison of one quantity to one unit of another quantity. The slope of that relationship is that same comparison.   **9.** Sample answer: The equation $y = x$ can be written as $y = 1x$ because $1 \cdot x = x$. So, the slope is 1, not 0.

Lesson 4-2 Slope of a Line, Practice Pages 203–204

1. $\frac{1}{4}$ or 0.25   **3.** $-\frac{4}{3}$ or $-1.\overline{3}$ **5.** 0
7. C

x	3	5	6	8
y	8	0	-4	-12

9. Sample answers: $(0, 0)$, $(5, -2)$, $(10, -4)$
11. Sample answer: The student did not subtract the y-coordinates and x-coordinates. The numerator should be $-4 - 8$ and the denominator should be $2 - (-3)$. The slope is $\frac{-12}{5}$.

Lesson 4-3 Similar Triangles and Slope, Practice Pages 211–212

1. The slope of segment *RT* is $\frac{1}{2}$ or 0.5. The slope of segment *TV* is $\frac{1}{2}$ or 0.5. The slopes of each segment are equal. **3.** Triangles *AHE*, *BID*, and *CGF*. The slope of each is $\frac{1}{2}$ or 0.5. **5.** The slope of the line is negative. The slopes of each triangle are the same because they lie on the same line. **7.** Sample slope triangles shown. The slope of line *m* is $-\frac{3}{2}$ or −1.5. The slope of line *n* is $\frac{2}{3}$ or $0.\overline{6}$. The slopes of perpendicular lines are opposite reciprocals.

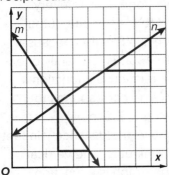

9. Sample answer: Because $\frac{8}{2}$ simplifies to 4, the triangles are similar slope triangles on the same line. Therefore, the slope of the line is always 4. **11.** No; Sample answer: The placement of the slope triangles does not matter because even if the triangle is placed above or below the line, the ratio of the vertical side to the horizontal side will always be the same as the slope of the line.

Lesson 4-4 Direct Variation, Practice Pages 223–224

1. $y = 8x$; The constant of variation is 8. This means that the cost per ticket is $8. **3.** $y = 2.55x$; The constant of variation is 2.55. This means that the cost per ream is $2.55. **5.** $y = 62.5x$; The constant of variation is 62.5. This means that the bus travels 62.5 miles per hour. **7.** $y = 4x$; 26 gal **9.** It is not a direct variation relationship. Sample answer: The ratio between the *y*-values and *x*-values is not constant. The points form a line, but the line does not pass through the origin. **11.** Sample answer: The constant of variation in a direct variation equation is the constant multiplied by *x*. The unit rate is also the constant multiplied by *x*.

Lesson 4-5 Slope-Intercept Form, Practice Pages 235–236

1. slope: $\frac{1}{2}$; *y*-intercept: −5 **3.** $y = -\frac{1}{3}x + 4$ **5.** $y = 4x - 2$ **7.** $y = 2x - 3$ **9.** $y = 30x + 30$ **11.** Coral Snorkeling; $5 less **13.** Sample answer: $y = -3$ **15.** Sample answer: rise = 18, run = 5

Lesson 4-6 Graph Linear Equations, Practice Pages 245–246

1.

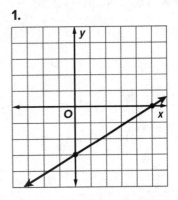

3.

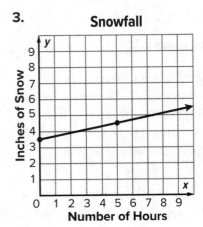

Snowfall

5.

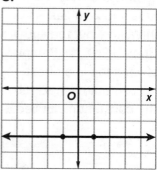

11.

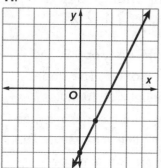

7. The slope of the line is negative. The equation of the line is $y = -\frac{7}{3}x - 7$.

9. Sample answer: Vertical lines will never have a y-intercept, unless the vertical line is the y-axis. So, there will be no b. Vertical lines have an undefined slope, so they will not have an m. **11.** Sample answer: If the slope is used to start graphing an equation, then the slope is comparing the first point plotted to the origin, and not to the y-intercept.

Module 4 Review Pages 249–250

1A.

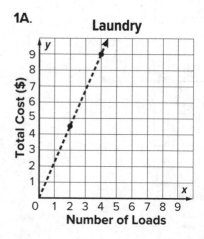

1B. Sample answer: The unit rate is represented by the point (1, 2.25), which lies on the line. **3.** B **5.** The ratio of the rise to the run of each triangle is the same. The smaller triangle and the larger triangle shown are similar. The slope of the line is −2.

7. $44.10 **9.** $y = -4x + 2$; I used the slope formula to find $m = -4$. The y-intercept was given.

Lesson 5-1 Identify Functions, Practice Pages 261–262

1. The relation is a function because every input value is mapped to exactly one output value. **3.** The relation is a function because every input value corresponds to exactly one output value. **5.** The relation is a function because every input value corresponds to exactly one output value. **7.** Sample answer: This graph shows the distance y in feet that a snail travels in x minutes. The vertical line test shows that the graph is a function because for every value of x, there is only one value for y.

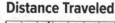
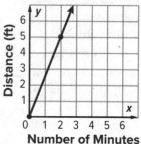

9. Sample answer: I could use a mapping diagram. If more than one arrow points from an input value to the output values, I know it is not a function.

Lesson 5-2 Function Tables, Practice Pages 271–272

1.

Input, x	Output, y
−5	−20.5
0	−8
5	4.5
10	17

3.

Input, x	Output, y
−2	2
2	4
6	6
10	8

5.

Input, x	Output, y
−2	5
−1	3
0	1
1	−1

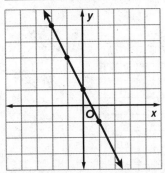

7. A giraffe is $4\frac{1}{2}$ feet taller in month 11 than in month 2. **9.** Sample answer: Graphs represent all solutions by using lines to show that solutions continue on in both directions. Equations represent all of the solutions of a function by using variables to represent any number. **11.** Sample answer: Kai switched the input and the output in the table. The values should be:

x	y
1	3
2	6
3	9

Lesson 5-3 Construct Linear Functions, Practice Pages 283–284

1. The rate of change is 8, so the hourly rate is $8. The value for y when $x = 0$ is 20, so the initial fee is $20; $y = 8x + 20$ **3.** The rate of change is 3, so the hourly cost is $3. The initial value is 3.5, so the skate rental fee is $3.50; $y = 3x + 3.50$ **5.** $22,000 **7.** 0; Sample answer: On a horizontal line, the y-coordinate never changes; so the rate of change is 0. **9.** Both are correct; Sample answer: The properties of operations show that both are correct.

Lesson 5-4 Compare Functions, Practice Pages 291–292

1. The function for Company A has an initial value of 500, while Company B has an initial value of 350. Company A has the greater initial value. The function for Company A has a rate of change of 12.5, while Company B has a rate of change of 15, so Company B has the greater rate of change; $80 **3.** The function for Lorraine has an initial value of 2.4. The function for Chila has an initial value of 1.7, so Lorraine started from farther away. The function for Lorraine has a rate of change of −0.2, while Chila also has a rate of change of −0.2, so the rates of change are the same. After 7 minutes, Lorraine is 1 mile from school and Chila is 0.3 mile from school. **5.** Both functions have the same rate of change, but different initial values. The graphs of the functions are parallel lines that will never intersect. **7.** sometimes; Sample answer: If the function with the greater rate of change has a lesser initial value, it will start with a lower output value. As the input values increase, the function with the greater rate of change will eventually have a greater output value.

Lesson 5-5 Nonlinear Functions, Practice Pages 303–304

1. The function is linear because the graph is a straight line and has a constant rate of change. **3.** The function is nonlinear because on the graph of the function, the ordered pairs (side length, area) do not lie on a straight line. **5.** The rates of change are the same, so the function is linear. **7.** The equation cannot be written in the form $y = mx + b$, so it is nonlinear. **9.** The relationship between the amount of money Catalina saves and the number of months is linear. The relationship between the amount of money Terri saves and the number of months is nonlinear. See students' arguments. **11.** Sample answer: Although it is a straight line, the graph is not a function because there is more than one value of y where $x = -3$. So, the graph is not a linear function. **13.** true; Sample answer: A non-vertical straight line is always a function, but a function does not have to be a straight line. A function can be curved.

Lesson 5-6 Qualitative Graphs, Practice Pages 311–312

1. Sample answer: Wesley ran in a direction away from home, and then sped up as he continued away. He then slowed down while still continuing away from home. Finally, he headed back in the direction of home at a steady pace until he reached home.

3. Sample answer: The graph is nonlinear. The graph is increasing when his heart rate is increasing. The graph is decreasing when his heart rate is decreasing.

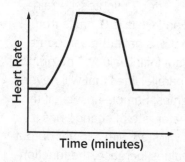

5. Sample answer: The team digs at a constant rate, takes a break for lunch, and then continues digging at a slower constant rate. **7.** Graph A; Sample answer: The graph of this relationship should increase at a constant rate, and then remain level. Only Graph A does this.

Module 5 Review Pages 315–316

1.

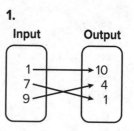

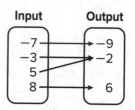

x	−8	−2	0	3
y	1	2	3	4

3.

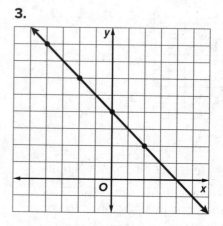

5A. B **5B.** The initial value is 40, so the flat delivery fee for the mulch purchase is $40.
$y = 35x + 40$

7.

	Linear	Nonlinear
$y = 2x - 9$	X	
$y = \frac{5}{x}$		X
$y = 3x^2$		X
$4x + y = 7$	X	
$y = \frac{x}{2}$	X	
$y = \sqrt{x+3}$		X

Lesson 6-1 Solve Systems of Equations by Graphing, Practice Pages 329–330

1. $(-2, 2)$

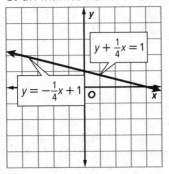

3. an infinite number of solutions

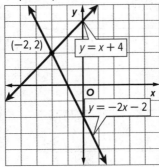

5.

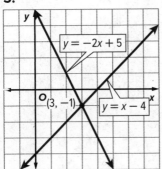

7. Kaylee drove 2 hours and Sophia drove 3 hours.

Driving a Scooter

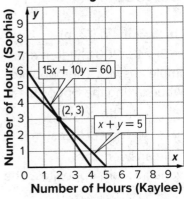

9. Sample answer: The point $(-1, -5)$ does not lie on the line of either equation.

Lesson 6-2 Determine Number of Solutions, Practice Pages 339–340

1. no solution   **3.** an infinite number of solutions   **5.** no solution   **7.** does intersect **9.** does intersect   **11.** The friends pass through a common coordinate.   **13.** Sample answer: Since $-7x + 3y$ cannot simultaneously be -5 and 9, there is no solution.   **15.** Sample answer: Perpendicular lines have slopes that are the negative reciprocal of one another.

Lesson 6-3 Solve Systems of Equations by Substitution, Practice Pages 349–350

1. $(2, -12)$ **3.** an infinite number of solutions

5. $(-1, -10)$ **7.** $(10, 9)$ **9.** no solution

11. $(-1, -6)$ **13.** The width is 7.2 feet and the length is 11.8 feet.   **15.** Sample answer: The equation $-1 = -1$ is always a true statement, so the system has infinitely many solutions.   **17.** sometimes; Sample answer: When solving a system of equations by substitution, one of the equations needs to be solved for a variable. So one of the equations may not necessarily be written in slope-intercept form.

Lesson 6-4 Solve Systems of Equations by Elimination, Practice Pages 361–362

1. $(2, 9)$ **3.** $(2, 1)$ **5.** $(9, -1)$ **7.** $(12, 5)$
9. $(2, -2)$ **11.** $(-1, -3)$ **13.** 900 children and 350 adults   **15.** Sample answer: The opposite of 4 was not used when taking the opposite of the equation $6x - y = 4$. The correct equation is $-6x + y = -4$. The correct sum is $2x = -14$.   **17.** Sample answer: The coefficients of the variable x are opposites. The value of a is -5.

Lesson 6-5 Write and Solve Systems of Equations, Practice Pages 373–374

1. Sample answer: $x + y = 20.5$ and $x - y = 6.5$; $(13.5, 7)$; The two numbers are 13.5 and 7.   **3.** Sample answer: $24x + 6y = 144$ and $18x + 3y = 90$; $(3, 12)$; It costs \$3 for each flower and \$12 for each bush.   **5.** 5 hours; \$75

7.

Week	Number of Days Ran	Number of Days Cycled	Total Number of Miles
A	2	3	40
B	3	4	55
C	4	2	40

9. Sample answer: $x + y = 12$ and $0.15x + 0.3y = 3$; $(4, 8)$; 4 liters of solution A and 8 liters of solution B

Module 6 Review Pages 377–378

1A.

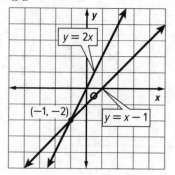

1B. $(-1, -2)$

3.

Systems of Equations	Number of Solutions		
	none	one	infinite
$y = \frac{3}{5}x + 1$ $5y = 3x + 5$			X
$y = 3 - x$ $y = -2x$		X	
$y = 2x + 3$ $4y = 8x - 1$	X		
$2x + y = 4$ $x - y = 1$		X	

5. C   **7.** A   **9A.** $y = 10x + 20$; $y = 15x$
9B. $(4, 60)$